# Imagining the Future City:
# London 2062

## Edited by
## Sarah Bell and James Paskins

]u[

ubiquity press
London

Published by
Ubiquity Press Ltd.
Gordon House
29 Gordon Square
London WC1H 0PP
www.ubiquitypress.com

First published 2013

Cover Illustrations by Edward Manley
Front: 'Clusters of Activity in Minicab Journeys across London'
Back: 'Mapping Minicab Flows across London'

Printed in the UK by Lightning Source Ltd.

ISBN (paperback): 978-1-909188-18-1
ISBN (EPUB): 978-1-909188-19-8
ISBN (PDF): 978-1-909188-20-4

DOI: http://dx.doi.org/10.5334/bag

Suggested citation:
Bell, S and Paskins, J (eds.) 2013 *Imagining the Future City: London 2062*. London:
Ubiquity Press. DOI: http://dx.doi.org/10.5334/bag

A minor correction was made to this book soon after publication. On p.77, the
sentence "When refurbishing, covering up solid walls with insulation can significantly
increase the available thermal mass…", was amended to "When refurbishing, covering
up solid walls with insulation can significantly decrease the available thermal mass…".
This error was present for a very short time and has been corrected in all formats of
the publication.

To read the online open access version of this
book, either visit http://dx.doi.org/10.5334/bag
or scan this QR code with your mobile device:

# Contents

# Images

Shad Thames (Mark Tewdwr-Jones)

Counter Street (James Paskins)

Upper Thames Street (Mark Tewdwr-Jones)

Computer (Sarah Bell)

Wetland Public Park (Charlotte Reynolds)

Recycling (Sarah Bell)

Red Pin (Sarah Bell)

Liverpool Street (Robin Hickman)

View from St Paul's (James Paskins)

Housing and Urban Dairy Farm (Michael Pugh)

Grey (Sarah Bell)

Hydro-London (Tse-Hui Teh)

Producing Food (Sarah Bell)

Boarded-Up Building (James Paskins)

Piccadilly Circus (James Paskins)

London Underground (James Paskins)

Surrey Row, Southwark (Mark Tewdwr-Jones)

In Our Hands (Nicole Hunter)

Paranoia House (Arthur Kay)

River Thames (Mark Tewdwr-Jones)

Oxford Street (James Paskins)

New Change Shopping Centre and Offices (Mark Tewdwr-Jones)

Lea Valley, Hackney (Matthew Gandy)

Tube Poster (Liron Schur)

# Acknowledgements

This book owes much to the robust, interdisciplinary community of urban scholars, researchers, practitioners and activists at UCL and in our London networks. It is the outcome of the London 2062 project which ran at UCL between 2010 and 2012. Many of the participants in the London 2062 seminars, symposia and conversations have contributed chapters to the book, but many more people made presentations and took part in lively discussion and debate. The book as a whole and individual chapters have benefited as a result.

Future of London was a key partner in a series of workshops looking at the future of London's housing, economy, energy and transport. Their members and associates provided useful contributions to the seminars and the book, and were very valuable contributors to discussion of key issues, bringing on the ground experience to academic reflection and analysis.

David Price, Vice-Provost for Research at UCL, instigated the London 2062 project as part of the Research Grand Challenge of Sustainable Cities. His vision was to bring the breadth of UCL's urban research to bear on the future of our own city. The project received support from key staff in his office, including Marianne Knight, Ian Scott and Nicholas Tyndale.

Mark Tewdwr-Jones was co-convenor of the London 2062 project until his move to Newcastle University in 2012, including early editorial work on the book. The project would not have been as well structured and informed, nor as much fun, without him.

The team at Ubiquity Press have been patient and professional in encouraging and supporting this slightly quirky book.

Rachel Cooper and Alvise Simondetti provided valuable feedback as peer reviewers of the draft manuscript. Their comments led to improvements the manuscript but they are not responsible for its residual flaws.

Charlotte Barrow, Research Assistant in Sustainable Cities, was utterly invaluable in getting us over the line.

# Contributors

**Michael Batty CBE** is Professor of Planning at UCL's Centre for Advanced Spatial Analysis (CASA). His current research is focused on new statistical models of cities with an emphasis on scaling, city size and morphology. He has written a series of books on these topics, the latest of which *The New Science of Cities* is to be published in late 2013 by MIT Press. He is Editor of the journal *Environment and Planning B*. He is Fellow of the Royal Society, the British Academy and the Royal Society of Arts.

**Matthew Beaumont** is a Senior Lecturer in the Department of English at UCL, and a co-director of the Urban Laboratory. He is the author of *Utopia Ltd.: Ideologies of Social Dreaming in England, 1870-1900* (2005) and *The Spectre of Utopia: Utopian and Science Fictions at the Fin de Siècle* (2012). Among other essay collections, he has co-edited *Restless Cities* (2010) with Gregory Dart. He is currently writing *Nightwalking: A History*.

**Sarah Bell** is Senior Lecturer at UCL in the Department of Civil, Environmental and Geomatic Engineering, and was co-convenor of the London 2062 Project. Her research interests lie in the relationships between engineering, technology and society as they impact on sustainability, particularly in relation to water systems. She is a Chartered Engineer and holds a PhD in Sustainability and Technology Policy from Murdoch University, Western Australia. She is a co-director of the Urban Laboratory and previous co-director of the Environment Institute.

**Robert Biel** teaches political ecology and urban agriculture at the UCL Development Planning Unit. In his published work, including his recent book *The Entropy of Capitalism* (2012), he applies general systems theory to an understanding of the current crisis and its possible solutions. He helped to run the ABUNDANCE programme introducing food growing in low income housing estates in partnership with Transition Town Brixton, and recently presented to Parliament on food futures. He participates in the allotment movement and actively experiments with sustainable food-growing techniques.

**Simon Cavanagh** is a Project Manager for a national housing charity. Working as part of their London Development team, he is currently leading on a multi-departmental approach to provide affordable homes in Westminster. Previous projects have included an award-winning regeneration of four estates in Hackney and the provision of sheltered housing in London and the South-

East. As part of the charity's commitment to younger people, he is keen to promote the benefits of working for a sector that will be key to ensuring London stays attainable for all sectors of society.

**Brian Collins CBE** is Professor of Engineering Policy at UCL and Fellow of the Royal Academy of Engineering. Between 2006 and 2011 he was Chief Scientific Advisor to the Department for Transport (DfT), the Department for Business innovation and Skills (BIS) and Department for Business, Enterprise and Regulatory Reform (BERR). He chaired the Engineering and Interdependency Expert Group for Infrastructure UK. He has served as Vice President of the Institution of Electrical Engineers and was Chairman of the Informatics Division. By presidential invitation, he is a Fellow of the Institution of Civil Engineers, in addition to being a Fellow of the Institute of Physics.

**Hannah Dalgleish** works at the London Borough of Hackney in the council's Regeneration Delivery team and is currently managing the delivery of a £5.3 million town centre regeneration programme in Hackney Central.

**David Fell** is co-founder of the research and strategy consultancy Brook Lyndhurst. He is an economist with more than 25 years research, scenario-planning and strategy experience for clients in the government, private and not-for-profit sectors. He has a particular interest in London: he has researched the capital's 'green economy', M25 property markets and the supply of and demand for skills in the Thames Gateway. He was a founding Commissioner on the London Sustainable Development Commission; and works as part of Just Space to support community and third-sector organisations to make fair representation within London's strategic spatial planning processes.

**Bob Fiddik** has worked on energy issues within local government for 19 years. This work has covered energy procurement, energy management, developing and delivering energy efficiency programmes for private and social sector homes, energy within planning policy and major regeneration schemes. This work has given him insight into the impact that national energy policy has on work on the ground. Bob has also worked on three major projects to establish district heat networks in London - at the Elephant and Castle, the 'South East London Combined Heat and Power' energy from waste plant and central Croydon.

**Matthew Gandy** was born in Islington, North London. He is Professor of Geography at UCL and was Director of the Urban Laboratory from 2006-11. His publications include *Concrete and clay: reworking nature in New York City* (2002), *Hydropolis* (2006) and *Urban Constellations* (2011), along with articles in *New Left Review, IJURR, Society and Space* and many other journals. He is also actively involved in local issues in Hackney, east London, and is a member of the Hackney Biodiversity Partnership. He has been a visiting professor at several universities including Columbia, Newcastle, UCLA, the Humboldt University, Berlin and the Technical University, Berlin. He is currently completing three book manuscripts: *The fabric of space: water, modernity, and the urban imagination* (for the MIT Press), *Moth* (for the Reaktion animal series), and a co-edited collection *The acoustic city* (for Jovis). From 2013 to 2015 he will be a senior research fellow of the Gerda Henkel Foundation at the University of the Arts, Berlin.

**Christine Hawley CBE** is Professor of Architectural Studies and Chair and Director of Design at the Bartlett School of Architecture. From 1999-2009 she was Dean of the Bartlett Faculty of the Built Environment, a multi-disciplinary group of over 200 academics. During this period the Faculty doubled its income to £18 million and achieved the highest rating for a single faculty submission in the 2007 RAE. She runs an architectural practice, has built social housing in Europe and Japan and developed typologies adapted to a range of different environmental conditions. Her work has received numerous awards and has been published in all the major international journals. She has recently received first prize for a multi-million pound cultural development in southwest China. She is advisor to a wide range of industry, government and academic bodies.

**Michelle Hegarty** is a housing and regeneration professional with over fifteen years of experience in delivering affordable housing solutions in London. She has been involved in developing and delivering housing projects, programmes and policy in the capital and has worked in a number of roles across central, local and regional government.

**Robin Hickman** is a Senior Lecturer at the Bartlett School of Planning, UCL and a Visiting Research Associate at the Transport Studies Unit, University of Oxford. He is Director of the MSc in Transport and City Planning. He previously worked in consultancy as an Associate Director and led on transport research at Halcrow (2004-2011); was a Research Fellow at the TSU, University of Oxford (2009-2011); and worked at Llewelyn Davies (1999-2004). He is a specialist in transport and climate change issues, urban structure and travel, integrated transport and urban planning strategies, and the management of multi-disciplinary projects.

**Nicole Hunter** is an Australian freelance illustrator and designer. She studied graphic design in Melbourne at the Royal Melbourne Institute of Technology before spending many years exploring the world. She found paradise, and settled in in tropical Cairns, Far North Queensland. There she works at the Australian Conservation Foundation, and creates pictures, usually with an environmental focus. She has worked on community murals, books and newsletters for food co-ops, environment groups, a circus, businesses and local council. She is currently working on her first children's book.

**Jennifer Johnson** is Programme and Research Lead for Future of London. She holds an MSc in International Planning from the Bartlett School of Planning at UCL, where she volunteered through the Just Space Network to provide planning assistance to community groups in London. She also has a BA in International Development Studies and Geography from McGill University. Previously her work has had a strong focus on community engagement while campaigning for human rights and environmental organisations including Amnesty International Canada, Oxfam Canada, and OneChange. Jennifer has a strong and growing interest in sustainable food systems, having spent two months learning about permaculture in the heart of Cuba.

**Arthur Kay** is an award-winning designer and entrepreneur, and graduate of the Bartlett School of Architecture, UCL. The relationship between design and the human psyche has always fascinated Arthur, and, in projects such as 'Paranoia House' and 'The Psychoanalyst's Kitchen' he has looked to explore this notion within an architectural framework. Alongside his work as a freelance designer and writer, Arthur founded the innovative, cradle-to-cradle green energy company bio-bean.

**Justin Kurland** is a PhD candidate at the UCL Department of Security and Crime Science. His research interests include security and crime-related problems at stadia events, the development of spatial and temporal analytical techniques for crime reduction and agent-based crime simulation.

**Ed Manley** is a Research Associate at UCL's Centre for Advanced Spatial Analysis (CASA) where he is working on the Mechanicity project to develop new models for understanding the morphology of cities. His research focuses on the use of large-scale datasets for the improved understanding and prediction of travel behaviour within the urban realm. In his current work, he is examining the use of Oyster Card data for the analysis, prediction and mitigation of disruption on the London Underground.

**Pablo Mateos** is Honorary Lecturer in Human Geography at the Department of Geography, UCL, and Associate Professor at the Centre for Research and Advanced Studies in Social Anthropology (CIESAS) in Guadalajara, Mexico. At UCL he is a member of the Migration Research Unit (MRU), an associate of the Centre for Advanced Spatial Analysis (CASA), and external fellow at the Centre for Research and Analysis of Migration (CReAM). His research interests focus on investigating ethnicity, identity, migration and segregation in contemporary cities in the UK, Spain, US and Mexico.

**Richard Milton** is a Senior Research Associate at UCL's Centre for Advanced Spatial Analysis (CASA) where he is the key developer of web-based mapping systems, in particular the MapTube portal. He has worked on the Equator e-Science project in UCL Computer Science where he used GPS tracked sensors to measure environmental factors and display carbon monoxide levels on a 3D model of the city. At Criterion Software, as part of the Renderware 3D engine for the games industry, he wrote art tools and plugins for 3DStudio Max. Previously, at the UK Meteorological Office, he developed weather visualisation systems for various commercial and military customers

**Janice Morphet** is a Visiting Professor in the Bartlett School of Planning, UCL. She is currently working on infrastructure planning and sub-state governance in the UK including the role of the British Irish Council and Devolution. Janice has held senior posts in local and central government, was Head of the School of Planning and Landscape at Birmingham Polytechnic and on the Planning Committee of the London 2012 Olympic Games. Her recent books are *Modern Local Government* (2008), *Effective Practice in Spatial Planning* (2010) and *How Europe Shapes British Public Policy* (2013).

**George Myerson** is a free-lance writer and Visiting Senior Research Fellow in the Centre for Life-Writing Research at King's College London. He has published 18 books, ranging from philosophy and social theory to biography and humour, including *Nostradamus' World Cup Prophecies* (2002). In 2009 he published *Fighting for Football*, an account of the life of Woolwich Arsenal star Tim Coleman before, during and after World War I. His latest book is *A Private History of Happiness: moments of fulfilment across the centuries,* published in the USA by Blue Bridge Books and in the UK by Head of Zeus.

**Peter North** holds a BSc in Engineering and an MSc in Building Services Engineering. He is a Fellow of the Institution of Mechanical Engineers and a Chartered Engineer. His professional career has focused on the energy sector in design, development and implementation for both the private and public sectors. His experience ranges from nuclear and fossil fuel, to combined heat and power, district heating and renewables, including waste. Peter's work involves the commercial development of energy projects, which combines engineering and commercial development with financing and project structuring. Peter is currently a Senior Manager – Programme Delivery (Sustainable Energy), with the Greater London Authority.

**James Paskins** is Co-ordinator for the Grand Challenge of Sustainable Cities and the London Agenda at UCL, initiatives which encourage cross-disciplinary research into complex societal issues. He is a member of the British Psychological Society and a Fellow of the Royal Society of Arts. He holds a BSc in Psychology from the University of Westminster, and a PhD in Transport Studies from UCL. His research interests cross the boundaries between transport, urban sustainability, psychology, and accessibility research. He has extensive experience in undertaking and managing interdisciplinary research, including being programme manager for the EPSRC funded Bridging the Gaps: Sustainable Urban Spaces project at UCL and researcher on a number of EPSRC funded transport research projects.

**Rob Pearce** is an Associate Director at Renaisi, a leading Social Enterprise based in the East End of London. His background is in government, finance and business development. As the company's lead social entrepreneur Rob has responsibility for new business and designing and delivering new approaches to public services. He has overseen social business acquisitions as well as the development of joint ventures with the private sector. He has worked at local and national levels and across the public, private and third sectors. He is a qualified accountant and has an MA from Kings College London.

**Sofie Pelsmakers** is a Chartered Architect and environmental designer with more than a decade of hands-on experience designing, building and teaching sustainable architecture. She taught sustainability and environmental design and led a Masters programme in sustainable design at the University of East London. Sofie is currently a doctoral researcher at the UCL Energy Institute and co-founder of Architecture for Change, a not-for profit environmental building organisation. She is author of *The Environmental Design Pocketbook* (2012).

**Michael Pugh** is an architectural designer based in London. He worked at Rogers Stirk Harbour + Partners from 2012-13 on a high-density, mixed-use residential scheme in Monte Carlo. From 2011-12 Michael worked at Rick Mather Architects on a mixed-use scheme for the Grade II listed Centre Point complex, which provides a new public square in the centre of London. Having studied for his BSc in Architecture at Bartlett UCL, achieving a RIBA Bronze Medal nomination, Michael has developed interests that emphasise innovative, sustainable and socially-aware responses to architectural problems on varying scales.

**Mike Raco** is Professor of Urban Governance and Development in the Bartlett School of Planning, UCL. His background is in planning, geography, and urban studies. He has published widely on the topics of urban governance and regeneration, urban sustainability, urban communities, and the politics of urban economic development. Recent works include: *The Future of Sustainable Cities: Critical Reflections* (with John Flint, 2012); and *State-led Privatisation and the Demise of the Democratic State: Welfare Reform and Localism in an Era of Regulatory Capitalism* (2013). He formerly lectured at King's College London and the Universities of Reading and Glasgow.

**Jonathan Reades** is a Lecturer in Quantitative Human Geography at King's College London. His long-standing research interest in 'big data' and 'smart cities' is driven by his work in communications and transport technologies and their impact on human geography, including: access to opportunity and mobility; firm location, clustering, and growth; and our understanding of human behaviour interaction on a vast scale. He is an Honorary Research Associate at UCL's Centre for Advanced Spatial Analysis (CASA).

**Charlotte Reynolds** was awarded a first class honours degree and was shortlisted for the RIBA President's Bronze Medal after completing her BSc Architecture at the Bartlett, UCL in 2011. Receiving the RIBA Fitzroy Robinson drawing prize further demonstrated her dedication to drafting and model-making as fundamental tools informing her design process. Working part time as research assistant for Professor Christine Hawley at UCL, and for Matthew Springett Associates, has given her an insight into architectural academia and international exhibitions alongside gaining experience in a traditional office environment.

**Jennifer Robinson** is Professor of Human Geography at UCL. Her book, *Ordinary Cities* (2006) develops a post-colonial critique of urban studies, arguing for urban theorising which draws on the experiences of a wider range of cities around the globe. This project has been taken forward in her call to reinvent comparative urbanism for global urban studies. She has also published extensively on the history and contemporary politics of South African cities, including *The Power of Apartheid* (1996) and is currently working on transnational aspects of Johannesburg's policy making processes.

**Yvonne Rydin** is Professor of Planning, Environment and Public Policy in the Bartlett School of Planning, UCL. She has written widely on planning and urban governance, with a particular emphasis on how to progress towards the policy goal of urban sustainability. Her most recent books are: *Governing for Sustainable Urban Development* (2010); *The Purpose of Planning* (2011) and, a companion volume, *The Future of Planning* (Policy Press). In this latest book, she explores the implications of low economic growth for the UK planning system and its ability to achieve goals of social justice and environmental sustainability.

**Liron Schur** recently completed the MSc International Planning, at the Bartlett School of Planning, UCL, specializing in Sustainable Urban Development. His dissertation research trip to The Gambia in Sub-Saharan Africa, where he examined the relationship between sustainable transport and responsible tourism development, inspired his submission to the London 2062 student competition: a vision of a post-climate change London, and a subsequent reversal of economic fortunes between the Global North and South. He holds a BSc Computer Science from the University of Haifa, Israel and worked for over fifteen years as a software engineer for various companies, including five years at Microsoft. While living in Tel Aviv he was active in the 'Residents for Florentine' group, promoting public spaces in South Tel Aviv. Liron's recent work involves the integration of IT with urban and strategic transport planning, his lifelong passion.

**Theodoros Semertzidis** has an MSc in Urban Regeneration from the Bartlett School of Planning, UCL, an MSc in Environmental Engineering from the University of Manchester and a BEng in Civil Engineering from the University of Salford. He has worked in a variety of environments including the Public Power Corporation of Greece, Hug Engineering in Switzerland, and the Hellenic Greek Army, and is currently working for Schmid Energy Solutions in Switzerland. His main interests and responsibilities have been renewable energy sources and methodologies investigating their implementation, with his most recent work being on future scenarios.

**Aiden Sidebottom** is a lecturer at the UCL Department of Security and Crime Science. His research interests are evidence-based policing, situational crime prevention and crime prevention evaluation.

**Myfanwy Taylor** is a PhD Student at the UCL Department of Geography and Bartlett School of Planning. Her research aims to contribute to efforts underway to contest and develop alternatives to the dominant narratives of London's economy as being driven by financial and business services. She is also a researcher in urban health at LSE Cities, and was previously a policy officer at the Department for Communities and Local Government and the Cabinet Office. She is a member of the International Network of Urban Research and Action (INURA), Participatory Geographies Research Group and Stadtkolloquium.

**Tse-Hui Teh** is Lecturer in Urban Design and Planning at the Bartlett School of Planning, University College London. She is an architect and an urban designer with ten years of professional experience. Her main research interests are in the use of the actor-network theory co-evolutionary framework to understand the persistence and reconfiguration of the materiality of space; and to create new methods of public participation for the alteration and expansion of urban infrastructure. Her current research has a focus on concerns about the water-cycle, sanitation and well-being.

**Mark Tewdwr-Jones** is Professor of Town Planning and a member of the Global Urban Research Unit at Newcastle University and one of the UK's leading authorities on planning, land use, and urban development. Mark was educated at Cardiff and London Universities. He has worked in local government in Devon, and at a number of universities (UCL, Cardiff and Aberdeen). His research focuses on the politics and governance of planning, including spatial planning, land use planning futures, place-making, media representations of planning and planners, urban and regional development, governance and devolution, housing as second homes, and representations of planning and urban life. He has authored or edited fourteen books, including co-authoring the fifth edition of the classic *Urban and Regional Planning* with Peter Hall. He has advised ministers in UK Government, the Welsh Government and Scottish Government, on aspects of planning, and recently served as a lead expert for the Government Chief Scientist's Foresight project on Land Use Futures. He was elected an Academician of the Academy of Social Sciences in 2011.

**Jean Venables CBE** is a leading consultant in flood risk management and water level management, Chairman of Crane Environmental, Chief Executive of the Association of Drainage Authorities, a Non-Executive Director of HR Wallingford, and Chair of the Customer Challenge Group for Sutton and East Surrey Water Company. Jean was President of the Institution of Civil Engineers in 2008-9, the first woman to be President in the Institution's 190-year history.

**Jeremy Watson CBE** is responsible for Arup's strategy for Science & Technology. Until November 2012, he was Chief Scientific Adviser to the Department of Communities and local Government (DCLG). He is a Chartered Engineer, a Fellow of the Royal Academy of Engineering, Fellow of the Institution of Civil Engineers, Fellow of the IET and IET Vice-Chairman Elect, and Fellow of the Royal Society for the Arts. Jeremy is a former Board member of the UK Government Technology Strategy Board, and a Board member of the Institute for Sustainability. Jeremy is a member of the Engineering & Physical Sciences Research Council (EPSRC). He is a Visiting Professor at the Universities of Southampton and Sussex, and Professor at UCL in the Faculty of Engineering.

**Joanna Wilson** is the Director of Future of London, an independent not-for-profit policy network focused on the challenges facing urban regeneration, housing and economic development practitioners in London. She has an MSc in Sustainable Urbanism from the Bartlett School of Planning, UCL and an MA in Cultural Management from Northumbria University. Previously she worked at the Commission for Architecture and the Built Environment (CABE), providing client support for major capital programmes such as Building Schools for the Future and the Arts Capital programme. She has also worked as a freelance art event coordinator, and editorial assistant and staff writer for a-n The Artists Information Company. Jo is also a local food producer and activist, she trained in urban growing with the Hackney-based social enterprise Growing Communities, and is part of a cooperative that grows organic produce on small plots of land in Hackney.

# Introduction

Sarah Bell, James Paskins, Joanna Wilson and Jennifer Johnson

London is one of the world's great cities, hosting an extraordinary colocation and concentration of people, industries, political power, religions, finance, ideas and creativity. It is a city where you will find eight million people and more than 100 spoken languages. A city that is both ancient and modern: at the forefront of innovation in fields such as art, technology, finance and education, while reflecting nearly 2,000 years of history in its layout, institutions and buildings.

The next fifty years are likely to see the global population swell to nine billion people, who will be living under conditions of continuing economic, political, climatic, cultural and technological change. What will London be like in 2062? It is impossible to know. Trying to predict the future of such a dynamic city, in such uncertain times, would be foolhardy. And yet, many of the decisions taken by London's leaders and citizens now will have consequences far into the future. London will face significant challenges in the next half century and it is wise to be prepared.

This book is the outcome of the London 2062 Project, commissioned under the auspices of UCL's Research Grand Challenge of Sustainable Cities. The Grand Challenges[1] are an expression of UCL's commitment to bring our collective knowledge and wisdom to bear on some of the most complex problems facing the world. As international leaders in urban research, drawing together the trends shaping the future of our own city over the next fifty years seemed like an obvious contribution to inform public debate and decision making. It also presented a significant risk. Academic reputations are not built on bold speculation, but on careful, thorough analysis of data and thoughtful development of theory. A project addressing the future of London poses some serious academic problems.

The academic problem with the future is that we just do not have the data. We have some theories, but the most reliable of these deal with the things that we are least concerned about. Newto-

---

[1] The UCL Grand Challenges build on the university's accomplishment, expertise and commitment by encouraging researchers to think about how their work can impact upon global issues, bringing specialist expertise to bear on some of the world's key problems: Global Health, Sustainable Cities, Intercultural Interaction and Human Wellbeing. Find out more at www.ucl.ac.uk/grand-challenges.

**How to cite this book chapter:**
Bell, S *et al.* 2013. Introduction. In: Bell, S and Paskins, J. (eds.) *Imagining the Future City: London 2062.* Pp. 1-4. London: Ubiquity Press. DOI: http://dx.doi.org/10.5334/bag.intro

nian theory allows us to be effectively certain that acceleration due to gravity in London in 2062 will be approximately 9.8ms$^{-2}$, but micro-economic theory is less reliable in predicting how affordable housing will be. We have some numerical models of the how the world works, but these are either based in theory, or on statistical analysis of past events. They can help to identify possible futures and assign probabilities to various outcomes, but they work best when applied to biophysical phenomena, which is useful, but inadequate, when dealing with a complex and contested entity such as London.

The academic problem with London is that it does not conform to our disciplinary structures. Universities were not designed to deal with interdisciplinary challenges such as the future of a world city with 2,000 years of legacy. Research about London is spread across departments ranging from Engineering to English. While our disciplinary perspectives are necessarily incomplete, bringing them together to deliver a coherent set of meaningful forecasts is impossible. Engineering data and literary criticism cannot simply be merged. We must respect the frames of reference and levels of explanation offered by different schools, and take care not to lose the power of specific disciplinary analysis in our efforts to bridge the gaps between different systems of knowledge.

Academics are not alone in admitting difficulty in addressing the question of what London will be like in 2062. Our partners from professional, government and community organisations across London face a different set of challenges in dealing with the future. The people who make decisions, formulate policies, create plans, design systems and advocate for particular interests in the city are all working for a future they barely have time to contemplate. Faced with the everyday pressures of professional and political life, annual budgetary negotiations and five year election cycles, opportunities for long term thinking are rare. Opportunities to think about the future outside the particular details of a specific sector are practically unheard of.

Despite academic reticence and professional constraints, the future of our city and what it means for the choices currently facing leaders and citizens presents us with a series of questions we cannot hide from. How many people will live here? What will we eat? Will we have enough water? How will we achieve a low emissions transport system? Who will be in charge? Who will own our housing? The final answers to these and other questions are less important than the choices that emerge in thinking through the problems. Working through these issues across disciplines, professions and sectors is most valuable in provoking new thinking about decisions in the present, not in making specific predictions about the future.

Uncertainty aside, practitioners and policy-makers rely on projections as best guesses of future data to inform their decisions. The precariousness of this is highlighted in London's population projections: a recent upward revision has estimated London's population in 2030 at ten million people, up significantly from a previous prediction of nine million, which will now be reached in 2020. This translates to an additional 3.9 million households in the Capital, a third of which will be single occupancy households.

Housing, moving, employing, and providing public services for two million additional people within two decades will be no small feat. The pressure of demographic change is already being felt with a substantial increase in the number of school-age children and a corresponding shortfall of 118,000 primary and secondary school places as soon as 2016. The transport network is also feeling the strain as these ten million people could mean an extra 2.1 million daily trips – a key reason Crossrail 2 has already gone to public consultation. And perhaps nowhere is the pressure felt more than in the housing sector, where predictions suggest that the number of households will grow by around 36,000 per annum to 2033, in a context where new housing supply in London has averaged 24,582 per year in 2007-2012.

Amidst meeting demands such as these, there are also ambitions to improve quality of life in the city. Environmental sustainability now cuts across all sectors, as the Mayor of London strives to achieve a 60% reduction in carbon emissions by 2025 and is proactive in growing the green economy. Affordability, community engagement, segregation, and accessibility are also on the agenda.

The real challenge is not in fulfilling these ambitions, nor only in providing the infrastructure and services to underpin a burgeoning population. The careful balance of growth and quality, often in seeming competition with each other, present the real challenge for London to 2062 and beyond.

The timeframe of the London 2062 project provided a unique vehicle for opening up discussion and debate about the city and the forces that shape its future. Fifty years is far enough in the future that no-one could reasonably be held accountable for a false prediction, but it is less than a lifetime, fitting within the normal capacity of personal and professional memory and imagination. Some participants in the project had close to fifty years of professional experience working in London and urban research, others had nearly fifty years ahead of them. Fifty years is uncertain enough to free participants from the fear of being proved wrong, but firm enough to be certain that some of the decisions and actions taken now will still be remembered and their impacts still felt.

The project began in 2010 with a series of seminars with UCL staff and selected partners addressing the themes of London as a global city, a healthy city, a sustainable city and a thriving city. Participants were asked to consider 'where are we now?', 'how did we get here?', 'what might happen?', 'what can we do about it?' and 'what would London look like?'. Topics covered included population, energy, health, security, culture, heritage, housing, water, transport, governance, air quality, waste management, finance, food and flooding. The outcomes of the seminars were synthesised into a pamphlet which addressed the current state of London, and the forces shaping its resilience, sustainability and wellbeing[2].

The second phase of the project centred on a series of workshops co-convened with policy network Future of London[3]. These events linked more strongly to urban practitioners and policy makers in London, drawing on Future of London's members and networks involving local government, the Greater London Authority, Transport for London and other organisations. The Future of London workshops covered energy, housing, the economy and transport. We also held a competition for UCL students to present their ideas and visions in writing or images.

This book is an outcome of the project. The contributors were participants in the seminars, workshops and competition, and other members of the UCL community with important or unique perspectives on London. Core issues such as demography, transport, water and governance are addressed in longer chapters conforming to the conventions of academic writing. Shorter chapters provide key contributors with the opportunity to speculate about possible futures or key issues in a freer style, ranging from the polemical to the fictional. Images are an important feature of the book, illustrating points addressed in the text and making their own statements about London's future.

The book is structured into four sections – *Connections*, *Things*, *Power* and *Dreams*. *Connections* addresses London's role as a global and capital city, and the inter-relationships between different elements of the city itself in shaping its future. *Things* addresses the material aspects of London, including infrastructures, food and waste. *Power* covers issues of governance, the economy and housing. *Dreams* covers more speculative contributions, including formal scenario approaches to the future of London and individual imaginations of London in 2062.

The London 2062 process of enquiry resulted in a diverse, but inevitably incomplete, set of perspectives on London's future, opening new ideas and ways of thinking about the city. The project did not follow a formal scenarios based approach, but several authors develop their own scenarios for the future of London and one chapter specifically addresses well established scenarios and

---

[2] Background information related to the London 2062, including the pamphlet and videos of seminar presentations is available at: www.ucl.ac.uk/london-2062

[3] Future of London is an independent, not-for-profit policy network focused on the challenges facing urban regeneration, housing and economic development practitioners across the Capital. More information is available at: www.futureoflondon.org.uk

what they might mean for London. Climate change is not addressed as a separate issue, but is embedded in analysis of infrastructure, the built environment and governance, and is a feature of most of the personal and creative contributions. Several of the chapters review current government, industry and scientific forecasts, including those for demographics, transport, energy and water. A number of other issues of current concern to London's politicians, businesses and residents are dealt with by scattered references throughout several chapters, but are not specifically analysed in detail. These include airports, finance, religion, manufacturing and the influence of the US, Europe, China and other emerging economies. Some of the chapters contradict each other, as individual visions and opinions sit alongside more conventionally structured analysis and synthesis. The chapters are written in different styles, reflecting the multiplicity of voices, disciplines and positions.

*Imagining the Future City* is not a blueprint, nor is it a prophecy. It is a provocation to continuing dialogue and a prompt for Londoners to consider what is at stake. London's future cannot be predicted, foretold or decided in a book, however expert its authors. This book presents the visions and analysis of authors from a range of academic and professional disciplines, with the aim of opening up deliberation and debate about what London might be like in fifty years, and what that might mean for the choices we make now.

# Connections

London relies upon its connections. Without flows of food, water and energy it could not exist. Without its links to local, national and international markets, it would not thrive. Data connections are essential for communication within the city and with the rest of the world. Cultural connections allow creative and sometimes divisive exchange both here and abroad. Political connections allow the capital city of the United Kingdom, the largest city in Western Europe, to act as a global centre for policy, debate and diplomacy.

Two of the most important questions for the future of London are 'how many people will live here?' and 'who will they be?' The size and structure of London's population in 2062 will be a fundamental factor in decision making and a major determinant for quality of life. *Connections* uses demographic statistics and trends to tell a remarkable story about London's past, present and possible futures. In this section we are shown how a map of the London Underground can reveal stark differences in life expectancy across the city. We learn how the city's fertility has been affected by the changing role of women, and by patterns of migration. Inward and outward migration provides human connections that extends across the country, through Europe and across the rest of the world, and dramatically influences the size, diversity and age profile of the population. London is a city for the young and its attraction wains during child bearing years and old age; future growth rates and population profiles will depend on its relative attractiveness to people from different places at different stages of their lives.

Flows of people, culture, money and ideas connect London to other global cities, but may also underpin local and global divisions. Cities such as New York, Tokyo and Frankfurt, along with emerging global cities in Asia, Latin America and Africa, compete with London for capital, power and prestige. Choices taken about the future of London have repercussions for people, economics and environments in faraway places, just as changes elsewhere in the world will impact on London. Whilst the diversity of London's population is celebrated as a reflection of its global connections, it often plays out in stark divisions along ethnic, economic and social lines. Connection and division characterise London as a global city, and will present key challenges for citizens and political leaders in coming decades.

London's infrastructure provides vital connections that make the city work: energy, water, transport and communications systems are taken for granted when they work properly, and adapting and upgrading these networks is a key task for utility providers and governments. Technological innovation holds the promise to make infrastructure smarter and more connected to the everyday needs of Londoners and the strategic needs of the economy. Global and local environmental changes and constraints will shape London's built environment and infrastructure. The success of the city will be dependent on judicious design and decision making to ensure resilience to internal and external shocks, adaptation to underlying trends, and the sustainability of the resources on which the city depends.

# London's population

## Pablo Mateos

## Introduction

A city is defined by its population: the number of inhabitants, their demographic characteristics and geographical distribution. Indeed, demography is key for planning a city's future, since the current and future number of people per age, sex and ethnic group and geographical area of residence determines all dimensions of urban life, from economic, health, education, fiscal, housing or transport policies to many other socio-cultural aspects. However, forecasting a city's demographic future is an extremely difficult undertaking, as will be discussed in this chapter. It is an endeavour that requires 'a mirror to the past and the crystal ball to the future'. This chapter will dive into London's population historical facts, to then adventure into possible future trends over the coming decades.

London's population size during the last two centuries has experienced quite a bumpy ride. At the very first Census of Population, in 1801, just 1.09 million people were enumerated within the current geographical boundary of Greater London. Throughout the nineteenth century London´s population increased six-fold to 6.5 million in 1901. This trend continued up to the beginning of the Second World War, reaching the largest population size that London has ever had; 8.6 million in 1939. The war actually marked the outset of five decades of continuous population decline, as a result of post-war reconstruction, slum clearance, suburbanisation, the 'green belt' restriction on sprawl, de-industrialisation, a decline in fertility rates and smaller household sizes. As a result, in 1991 London's population had been reduced to 6.4 million people, less people than it had at the start of the 20th century. Since then, trends in population decline have reversed. Through processes of international migration, urban renewal, a sustained economic boom, and a recent rise in fertility rates, London has grown over the past two decades to reach 8.17 million in 2011, its former size in 1931.

This story summarised in a total people headcount actually hides a range of very different population dynamics that will be teased out throughout this chapter. The most classic of such dynamics is geographical differentiation, or a tale of two different Londons: Inner and Outer London. Figure 1 shows that the total population of Inner London boroughs actually peaked around 1911 at just above

**How to cite this book chapter:**
Mateos, P. 2013. London's Population. In: Bell, S and Paskins, J. (eds.) *Imagining the Future City: London 2062*. Pp. 7-21. London: Ubiquity Press. DOI: http://dx.doi.org/10.5334/bag.a

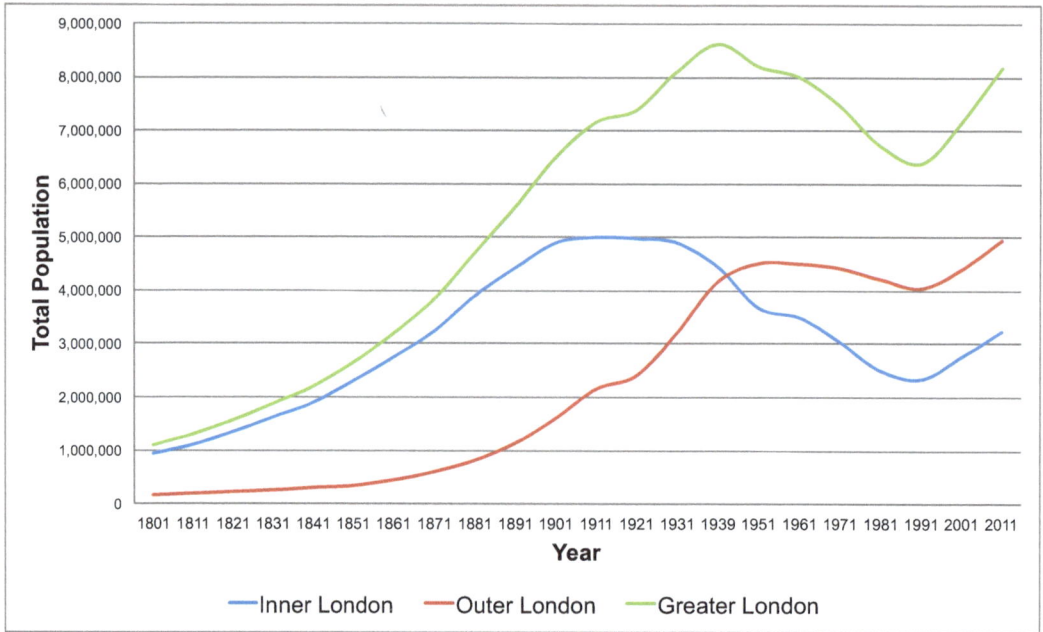

**Figure 1:** Inner and Outer London's total population (1801-2011) (Greater London Authority, 2012). The top line (Greater London) shows the combined total population of Inner and Outer London, which also appear as individual line graphs below it.

five million people, a point from which it continuously lost population throughout most of the 20th century, halving to 2.3 million in 1991. However, over the last two decades it has gained almost 0.9 million people, averaging 1.6% growth per year, reaching 3.2 million in 2011. Meanwhile, Outer London population only peaked in 1961 at 4.5 million, surpassing Inner London by one million at the time, and signalling the end of a long process of suburbanisation that started in the 1870s and 80s with the expansion of the suburban railways and the underground system. Between 1961 and 1991, Outer London lost half a million people, mostly due to de-industrialisation, but since then it has added another million people (a growth averaging 1% per year), reaching 4.9 million in 2011.

The main driver of population growth over the last two decades has been international migration, although since the late 2000s the largest component is actually natural change: the difference between births and deaths. This is due to London's extremely young population structure, compared to the rest of the UK. However, London's total population figure is far from being a stable mass of people, but instead reflects the net sum of a set of complex in and out flows bringing in vast amounts of young people in their 20s and 30s from the rest of the UK and the world, who typically have children in London and then emigrate outside London either as young families in their mid-to-late 30s, or at post-retirement ages. Such a demographic state of flux basically pertains to three types of events at the individual and household level; where and when people move, have children and die. This chapter reviews the factors and key trends behind such demographic events, and how they may shape up the peopling of London in 2062.

### Inequalities in life and death: natural change

The two obvious defining factors in a person's life are its beginning and its end, or a birth and a death. Indeed, where and when these two events occur, summed over everyone, have important consequences for the population size of any city, region or country. The difference between the

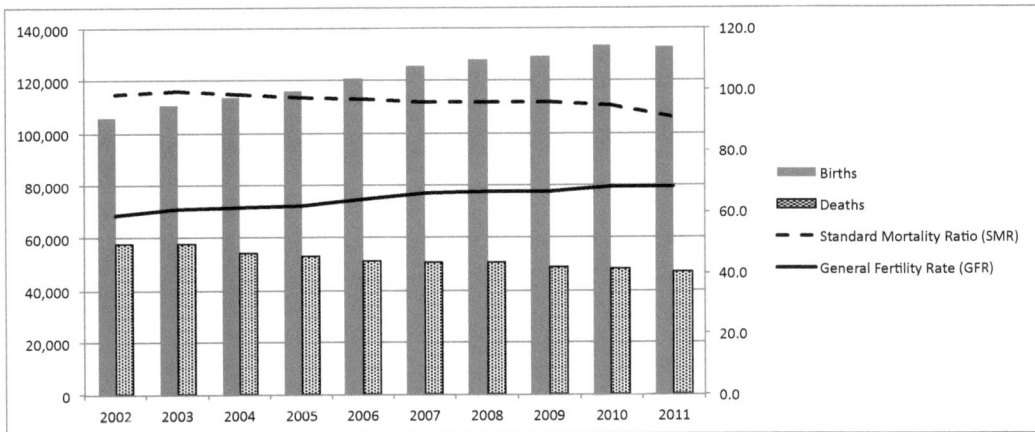

**Figure 2:** Evolution of natural change, fertility and mortality (2002-2011) (Greater London Authority, 2011; 2012).

dates of these two events, calculated over a person's life and averaged out across a whole population, gives us its life expectancy. Calculated over a city, region or country, the difference between the number of people being born and those dying over a particular time period (normally a year), gives us its natural change.

However, for most inhabitants of London, both life events rarely happen within Greater London's boundary, hence the life expectancy of Londoners is a summary of health conditions and demographic characteristics of populations across the globe. Although many people are indeed born in London, they are highly likely to emigrate as children before school age. At the other end of the life cycle, many Londoners are also likely to emigrate after retirement or reaching older ages, hence dying outside London despite having spent a lifetime there. This all means that figures on the life expectancy of Londoners are obviously problematic, since they are calculated from vital statistics (births and deaths registers) drawn within the Greater London boundary. This means that when talking about London's populace, we are never referring to a closed or static population, but one comprised of constantly changing membership. With this caveat in mind we will now review some of those vital statistics and life expectancy trends in London.

Because of London's imbalanced age-migration profiles, which will be discussed in the following section, London's natural change is widely positive. In other words, there are more children being born than people dying in any particular year within London's boundary. In 2011 natural change accounted for an increase of 86,158 people (Greater London Authority, 2011), the difference between 132,843 births minus 46,685 deaths. As indicated in Figure 2, this difference has been steadily increasing over the last decade. This is primarily a result of a surge in the number of births, derived from a growing and young population, but also because of a slight decrease in the number of deaths. The latter could be caused by selective emigration rates of older people with health issues that result in 'premature' mortality elsewhere, whilst boosting London's overall life expectancy. As a result of these trends, and because of a stabilization or reduction in net migration rates (the difference between immigration and emigration), natural change has become the main driver of population growth in London over the last five years (2007-2012).

## Fertility

Figure 2 also shows that such a steady increase in fertility is not only due to a larger and younger population, but also to an increase in relative fertility rates. The general fertility rate (GFR) used in the figure, is the annual number of live births per 1,000 women of childbearing age (aged 15 to 44). Between 2002 and 2011 the GFR has increased from 58.6 to 67.8, or nearly ten births per 1000 women of childbearing age. The Total Fertility Rate (TFR) is the number of live births per woman at the end of her reproductive life, if her childbearing at each age reflected current age-specific fertility rates. The TFR for London went from 1.29 children per woman in 2001 to 2.0 in 2010, after which it dropped to 1.84 in 2011 (Greater London Authority, 2012). Such increase in fertility rates is the combined result of several processes of population change: postponement, migration and ethnicity.

Postponement refers to the well-known trend of postponing childbearing to older ages. When an entire generation postpones the birth of a first child five to ten years compared to their parents' generation, this has a substantial impact on annual fertility rates when summed over the whole population. Such impact is known in demography as the *tempo* effect, or a transition between two different fertility regimes (Bongaarts & Feeney, 1998). If at the end of a woman's reproductive age, she and her cohort (those born in the same year) have had the same number of children as her mother did, there is no overall fertility decline (cohort fertility). However, motherhood postponement is likely to be noticed in any given time period. Part of the very low TFR rates in London in the late 1990s and early 2000s were actually caused by this transitory *tempo* effect, combined with emigration.

Furthermore, migration and ethnicity have also had an impact on London's fertility. Migrants tend to be younger and hence are more likely to have children than natives, which contributed to an increase in the absolute number of children born in London during the 2000s. Finally, some socioeconomic groups (the poorest and the very well off) as well as some ethnic minority groups, have slightly higher fertility rates than the average population. However, it has been demonstrated that ethnic minorities' fertility rates tend to level off with the general population very rapidly soon after migration (Coleman & Dubuc, 2010). London's national share of ethnic minorities, the very affluent and the very poor has increased over the last decade and this has probably had a small impact on its TFR over this period. In any case it is important to remember that many women do not actually spend their whole childbearing age in London. Therefore, it is difficult to extrapolate annual TFRs from actual childbearing behaviour of women who have not reached the end of their reproductive age and are currently residing in London.

Furthermore, fertility rates vary enormously across London. The TFR reaches 2.87 children per woman in Newham and the more deprived parts of Eastern London, while falling to just 1.18 in Westminster, a difference of more than a child 'and a half' per woman. Differences at smaller neighbourhood scales are probably much higher, although the TFR is not available below the borough (local authority) level.

## Mortality

Unequal mortality patterns across London have been a key characteristic of the city at least since John Graunt, probably the first demographer in history, studied them in the 17th century in his 'Bills of Mortality' (Graunt, 1665). One of the most commonly-used measures of mortality is the Standard Mortality Ratio (SMR). The SMR measures whether the population of an area has a higher or lower number of deaths than expected, based on the age profile of the population, compared to a national average or standard population. The SMR is expressed as a ratio with a base of 100. An SMR above 100 implies that there is 'excess mortality' whereas one below 100 suggests below-average mortality. London's SMR has declined from 98.5 in 2002 (i.e. close to the UK average of 100) to 90.9 in 2011, indicating a much lower mortality ratio than the overall UK population.

Life expectancy at birth in London in 2010 is 83.3 years for females and 79 for males (Office for National Statistics, 2011), slightly higher than for the whole of the UK; 82.3 and 78.2 respectively. This difference with the national average is attributed to entrenched wealth and health differences across the country, with London being more affluent and hence its population living slightly longer. However, this difference is also a consequence of selective outmigration of less healthy and older Londoners in middle and elderly ages, moving closer to their families or to affordable retirement destinations, and hence more likely to die 'earlier' and outside of London than more healthy Londoners who stay put. As a result, at age 65, life expectancy in London is 21.5 years for females and 18.7 for males. This means that those who have made it to 65 will live on average to 86.5 in the case of females and 83.7 in males. However, these average figures actually disguise startling geographical variations across London's unequal neighbourhoods. A commonly repeated illustration of such inequalities is that moving east on the underground along the Jubilee line, life expectancy between the boroughs of Westminster and Tower Hamlets drops at an average rate of 1 year of life per underground station (Atkinson, 2006). Figure 3 reproduces a popular map of such stark differences in life expectancy around each underground station in central London published by UCL Centre for Advanced Spatial Analysis (CASA) (Cheshire, 2012). Using this map's metrics to assign neighbourhood life expectancy (at birth) to each underground station, the starker and closest contrast happens to be on the Victoria line. Moving from Oxford Circus, with life expectancy of 96 years, just one stop to Warren Street, very near University College London (UCL), life expectancy drops to 79. A whole 17 years of life removed from populations neighbouring just 800 metres, or the same number of years as the difference between Japan and Bangladesh's life expectancies. Other stark contrasts appear in the inner-London section of the map reproduced in Figure 3 but many more are included in the full on-line version of this map, such as the clearly-indicated north south-divide.

Despite these favourable mortality trends, London's population is far from being considered healthy in terms of morbidity (prevalence of disease). Paralleling the aforementioned geographical disparities in life expectancy, some London boroughs actually present the starkest health inequalities in the country. Furthermore, London as a whole falls behind other large UK cities in many public health indicators. For example in low birth-weight babies, teenage conceptions, childhood obesity, childhood immunisations, chlamydia infection, HIV prevalence, incidence of tuberculosis, mental health, decayed teeth, or heart disease and strokes (Baker et al, 2012). Some of these 'less healthy' Londoners actually move out of London in middle age or after retirement, to perhaps find an early death somewhere else, hence contributing to 'improving' London's overall mortality figures. Despite their importance for people's quality of life, it is not the purpose of this chapter to discuss wider health and wellbeing issues, unless these have an effect on demographic trends.

## London´s revolving doors: migration

With respect to migration, London comprises a unique 'demographic laboratory', only found in a few other world cities. It is characterised by a very high population turn-over rate, a phenomenon we call 'London´s revolving doors'. Every year around 9% of its population moves into London while close to 7% leaves its territory. In 2011 alone the total number of in-movers and leavers added up to 1.3 million. Put in other terms, during the ten year period 2002-2011, a total of 7.3 million inflows and six million outflows took place (Greater London Authority Intelligence Unit, 2012). In a city of 8.1 million inhabitants in 2011, such a high population turnover rate has dramatic implications on a number of fronts. Evidently, some of these distinct flows will be generated by the same individuals coming and going, and thus being counted multiple times. However, it is very likely that a substantial amount of them are unique person-moves within the decade, meaning that a significant proportion of London´s population is replaced within a decade. The most striking consequence of these revolving doors is how fragile the overall balance of population actually is, and it is surprising how little it is discussed in public debates. Let us untangle what makes London´s doors revolve.

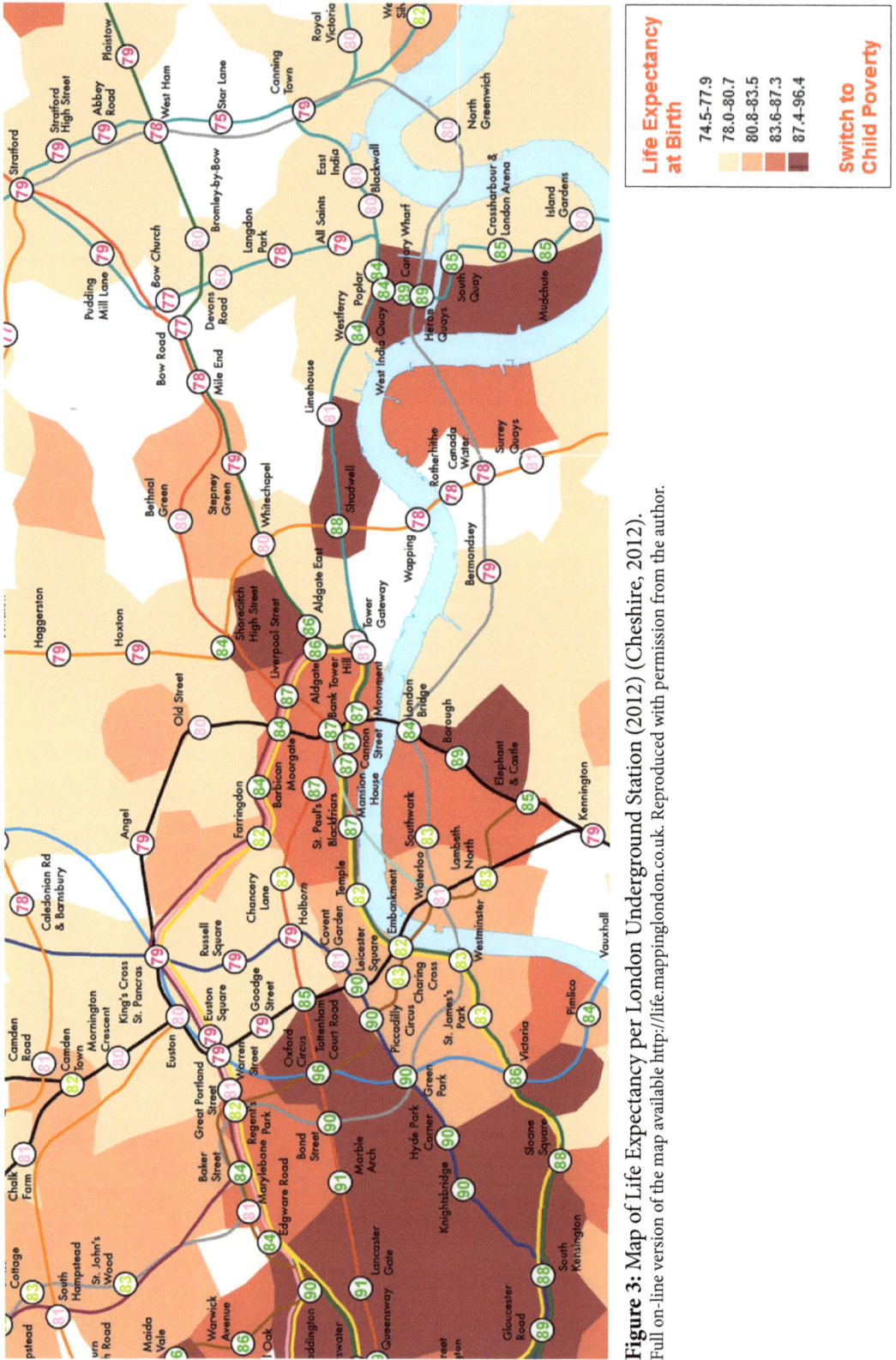

**Figure 3:** Map of Life Expectancy per London Underground Station (2012) (Cheshire, 2012).
Full on-line version of the map available http://life.mappinglondon.co.uk. Reproduced with permission from the author.

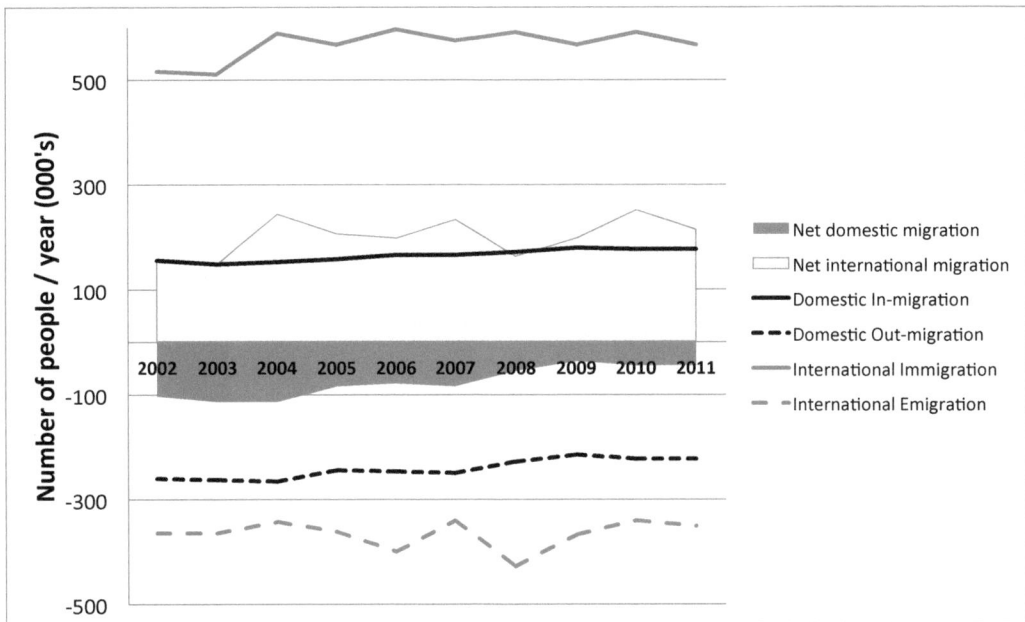

**Figure 4:** London migration flows (2002-2011) (Greater London Authority Intelligence Unit, 2012).

Migration is the third demographic driver in any population, together with births and deaths. Over the last decades London has been widely conceived as a city of immigrants, drawn both internationally and domestically. However, the absolute number of migrants added to any population (also termed 'stock') is necessarily the net difference of people moving in minus people moving out. This concept is termed *net migration*, and for a city like London it is actually calculated adding four distinct *gross flows*: two domestic migration flows (out- and in-migration from/ to the rest of the country), and two international flows (immigration and emigration from/to the rest of the world). The difference between the former pair is known as *net domestic migration* while that between the latter; *net international migration*. Despite the fact that these four flows are largely independent of one other, they together determine London's fragile population balance. Figure 4 shows how this balance has been sustained over the last decade, breaking down the four gross flows and the two net migration components for London. It is therefore very important to tease out the underlying factors driving each of these four *gross* flows in order to understand their impact on the two *net migration* components. The key explanatory driver of these migration flows is their age profiles.

In net terms, the key characteristic of London's migration system is a city that draws in massive numbers of people in their 20s without children, and tends to expel everyone else. Figure 5 shows the age profile of inflows and outflows in 2009. We observe two interesting phenomena that form part of the same 'revolving doors' balance. First, London as a whole loses population in all age groups except the 20-29 age group (the only one showing a positive net migration rate). Second, two age groups present a higher negative net migration rate: a) children aged 0-4, and their parents in the late 30s and early 40s, representing young families seeking more space and better schools before children reach school age b) people aged 60-64 and 75 and older, respectively representing near-retirement and later-life out-moves. Finally, if we calculate the ratio of outflows to inflows for each 5-year age group, we find that there are twice or more outflows than inflows between ages 0-14 and 55-75+. This fact clearly reveals London's 'expulsion' of less economically active populations. In other words, London acts primarily as a city for work, where all other aspects of the life-course are hard to sustain.

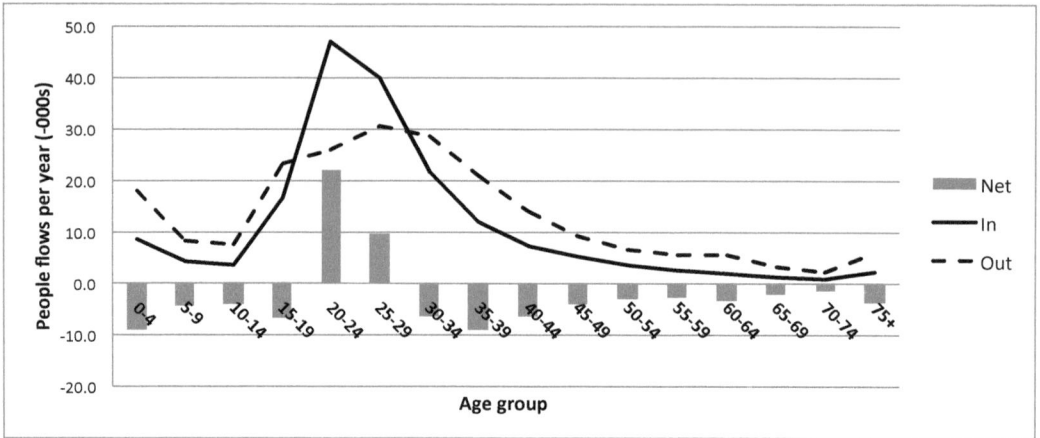

**Figure 5:** London migration flows by age group (2009) (Greater London Authority Intelligence Unit, 2012).

Geographically, these moves take place over a range of distances. Domestically, London net migration flows are positive with all UK sending-regions except the East, South East and South-west Government Office Regions (GORs). London actually loses population to these three GORs which are within commuting distance of London. Overall, London draws in large numbers of young people at university-entry or job-entry ages from all of the UK, who mostly leave London in their 30s or at retirement for the South-East regions. In Figure 4 we can clearly see that domestic net migration in London has been negative for the whole of the previous decade, although the absolute number has declined, especially since the outset of the recession in 2007-2008.

Internationally, the situation is too complex to summarise here, but we know from Census data that age profiles of immigrant inflows to London are slightly older than UK-born in-migrants (mostly in their late 20s to early 40s) and that they tend to emigrate to other countries and not to other UK regions. Furthermore, were it not for sustained positive international net migration flows averaging 200,000 people per year (2002-2011), London would have lost population very rapidly over the last decade. Figure 4 shows that net international migration is actually driven by fluctuations in emigration flows and not so much in immigration flows. Furthermore, unlike in other UK regions, immigration flows into London have largely been sustained at pre-recession levels. Emigration flows increased slightly in 2008, at the beginning of the recession, but have gone back to previous levels since then. Understanding the nature of such emigration and immigration flows' fluctuations is even more complex, since over 220 countries are involved and international migrants also include British nationals (those 'coming back' to the UK as well as native emigrants or naturalised migrants 'going back' to other countries).

The reasons behind decisions to move in and out of London depend not only on perceptions of labour opportunities, housing, or education prospects in London and the rest of the UK, but how these compare to those available worldwide. These factors are of course intersected with a person's and household's life-stage, family formation and housing cycles, as well as other socio-economic and cultural factors. Will London's revolving doors keep churning people at such high velocity over the next decades? What if one or several of these four types of migration flows changes drastically? Will the city be able to sustain its economic and cultural vibrancy without a vast incorporation of international immigrants arriving in their 20s and 30s who feed in and counterbalance the erosion in the rest of the demographic system? Will this fragile balance of demographic events that keeps London's population thriving be sustained in the future? These are crucial but mostly overlooked questions when talking about London´s population.

## How many Londoners in 2062? Forecasting population trends

Existing demographic forecasts produced by different agencies all predict that London's population will increase over the next decades, reaching between nine and ten million residents at some point between 2030 and the middle of the 21st century. However, these projections rely on a myriad of assumptions, most of them too difficult to predict over long periods of time. They have to do with the reasons behind decisions on when and where people decide to move or to have children, not just in London but in the rest of the UK and especially around the world. However, current population growth trends can revert quickly with just a slight change in one of the demographic elements sustaining London's overall population balance, as we have seen in the recent past.

Population projections are necessarily based upon assumptions about the future behaviour of each of the three demographic drivers: fertility, mortality and migration. Out of these three, the easiest to predict is future mortality rates. Improvements in life expectancy and morbidity rates have been fairly stable for most age groups. Furthermore, the majority of the base population to which these mortality rates will be applied has already been born. This means that it can be mechanically 'aged', one year every year, applying agreed mortality rates to each age group. However, fertility rates are much more difficult to predict, although not as problematic as migration rates. Future fertility rates are based on the predicted total fertility rate (TFR) broken down per age group (sometimes also by ethnic group). Therefore, fertility forecasts are based on two aspects: on the one hand, the projected age and gender population structure; and on the other, on predictions about decisions on when and how many children women in childbearing ages will have. The latter is extremely difficult to predict, since past fertility trends cannot be linearly projected into the future, because they concern a host of economic, cultural, lifestyle, and other social factors about the people now alive as well as others to come. Will childbearing postponement trends revert in the future? Will London sustain recent increases in the TFR close to replacement rates or will it go back to extremely low levels such as the 1.3 children per woman of the early 2000s? Will advances in assisted reproduction techniques allow older women to become mothers, and will the generalised use of these techniques substantially increase the already high number of multiple births in London? Demographers can only do some guesswork based on expert opinions and observe historic cycles, averaging out future trends into a single TFR for a range of years wrapping a bracket of confidence intervals around it.

As we have seen in this chapter, migration is the key driver in London´s population and does in fact control the dynamics of the other two demographic factors. Predicting whether London will keep attracting a large number of young people from the rest of the UK and the rest of world is more of an art than a science. Apart from maintaining the economic, entertainment and lifestyle attraction of the city, there needs to be enough affordable housing for everyone, a transport system able to cope with an increasing number of journeys, as well as other public infrastructure such as accessible schools and healthcare facilities wherever population increases. Given the slow pace at which such infrastructure is created, the lack of space and an aversion for higher population densities, it is difficult to foresee where an expanding number of Londoners will live. Furthermore, future political scenarios of the UK leaving the European Union or the single market, or if it stays, the prospect of further EU expansions, will definitely have an impact in London´s migration as we have seen in the 2000s. However, net migration is not only driven by people moving in, but also those moving out. If the current positive net migration rate for the 20-29 age bracket (seeFigure 5: London migration flows by age group (2009) (Greater London Authority Intelligence Unit, 2012). 5) is substantially eroded by a higher number of out-migrants in that group, the net migration figure for the whole of London might come close to a neutral or negative rate. In fact, various population projections seem to predict the latter scenario (see Figure 6). Finally, predictions on migration rates tend to be probabilistic, by including a set of future migration scenarios of high

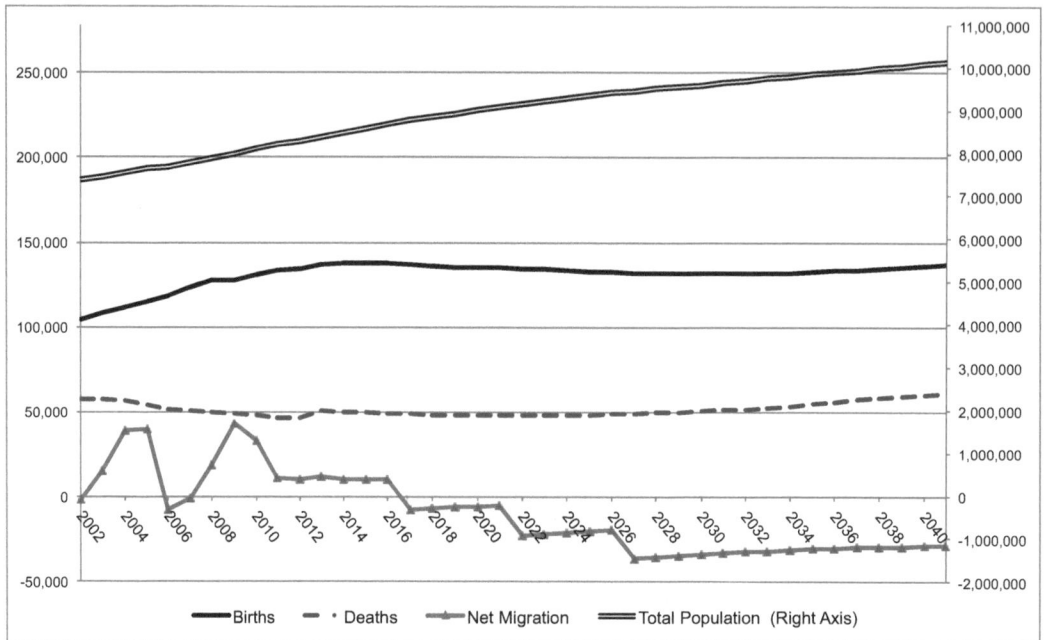

**Figure 6:** GLA London population projection 2002-2041 (Greater London Authority, 2012).
The three demographic drivers (births, deaths and migration) are represented on the left y-axis, while the resulting total population in the right y-axis.

and low migration phases, each of which is assigned a probability. Therefore, final population forecasts will have a buffer of confidence intervals delimited by such upper and lower brackets within set probability levels.

A more refined type of population forecasting model takes into account the overall housing supply available in the city each year, so that the total population projected for each period is constrained by it. These models need to add a further assumption on the future household and/ or dwelling size, in order to convert housing stock to total population. The number of persons per household in London increased from 2.3 in 2001 to 2.5 in 2011 (2.4 in in England and Wales), contravening a trend for smaller households in most of the developed world for the last decades. One plausible explanation is London's rapidly rising housing costs, which may be forcing more people to share dwellings or stay longer with parents, although this hypothesis needs testing with longitudinal data. In Figure 6 a housing-constrained forecast is shown, produced by the Greater London Authority (GLA) in 2012 for the period 2011-2041, as well as showing past population trends. In this forecast we see the population of London increasing to ten million people by 2041, despite the net migration rate being negative from 2017-2041. One of the reasons why net migration is negative in this forecast is precisely because the housing supply cannot be increased fast enough to cope with natural population growth. However, whether the timing of those births will happen before or after out-migration from London is a tricky question that brings us back full circle, showing the circularity and interdependence between the three demographic drivers.

Finally, most population forecasts are built with projections focussed on a 25-30 year horizon, roughly one generation ahead. Beyond 2041 and onto 2062, the science of predicting the population of London starts to become science fiction, since the number of unknowns outweighs what we can realistically control for. As we have seen in this chapter, a great share of the population of London turns over within a decade, and therefore very few Londoners are alive today who will still be residents in 2062. The domain of science fiction does indeed bring interesting scenarios on how London could look in the future. Assuming an extremely low fertility scenario (zero births) and high immigration, the movie *Children of Men* (directed by Alfonso Cuarón in 2006 and featuring Clive Owen, Julianne Moore and Michael Caine) presents an unsettling urban dystopia for London in 2027, reminding us of the power of demography when thinking about cities' futures.

## A diverse population

As a result of the aforementioned demographic processes, London is more 'ethno-culturally' diverse than any other Government Region or urban area in the rest of the UK, and probably than any other time in the city's history. The 2011 Census presents a unique opportunity to study ethnic and cultural diversity in the UK, since it comprises one of the few censuses in the world measuring a broad range of diversity dimensions (Mateos, 2014). As many as ten different Census questions are available relating to different aspects of these dimensions: country of birth, ethnic group, national identity, passport-citizenship/s held, religion, language/s most often spoken at home, English language proficiency, place of residence a year before the Census, year of arrival to country, and intended length of stay. Only some of the key trends in these variables will be summarised here, without cross-tabulating data between them (unless otherwise specified, all data in this section is taken from the Office for National Statistics, 2012). For more details on these questions for London, the reader is referred to the London Data Store resource (Office for National Statistics, 2012).

London is unquestionably the most ethnically-diverse area of the UK. In 2011 55% of London´s population was not White British (19.5% in England and Wales, hereafter E&W), rising from 40.2% in London in 2001. This share of the population is comprised of many different ethnic groups, and although 'Non-White' groups tend to get all the media and academic attention, actually 15% of London´s population belong to White groups different to White British (which itself includes all Whites with a UK national or identity adscription).

Looking just at immigrants, defined as those born outside the UK, these comprised 36.7% of London's population in 2011 (13.4% in E&W). If London was a country on its own, this would represent one of the highest relative migrant stocks in the world. Excluding city-states and small islands, only Israel, Jordan, Luxembourg and the United Arab Emirates have higher relative migration stocks than London. 48.2% of these immigrants had arrived in the UK within the ten years prior to the Census (2001-2011), while the rest have been living in the UK for longer. Therefore, roughly half of London's immigrants are recent migrants. Furthermore, 21% of Londoners have a non-British passport (7.4% in E&W), and 3.1% of Londoners have multiple passports (1.1% in E&W). The difference between the number of migrants born abroad and those without British passports is accounted for by naturalisations or acquisition of British citizenship through ancestors, which together add up to 15.7% of London's population (6% in E&W).

With respect to language, 26% of London's households have at least one person whose main language is not English (8.8% in E&W) and in 12.9% of households no one has English as their main language (4.3% in E&W). Beyond the Census, other data sources show even higher diversity trends, especially for the younger generations. For example, 40% of pupils in London schools speak a language at home other than English, covering a total of 322 languages (Von Ahn et al, 2010). If these large shares of the population retain their 'main languages', as well as learning English, London will indeed consolidate its role as a language hub in the world over the coming decades.

In terms of religious beliefs, Londoners are actually more religious than the rest of the country, since the percentage of people who declare 'no religion' is 20.7% compared to 25.1% in E&W. However, the share of Londoners who consider themselves Christian is just 48.4%, while it is 59.3% in E&W. Hence, a higher share of Non-Christian religions also indicates a much more diverse religious distribution. In descending frequency these are: Muslim (12.4%), Hindu (5%), Jewish (1.8%), Sikh (1.5%), and Buddhist (1%), while others are below 1% of the population. Of course the Christian label disguises a broad range of independent denominations that are not recorded separately in the Census (Anglican, Catholic, Orthodox, Evangelist and many others).

These broad-brush figures do not, however, reflect the nuanced and complex diversity of London's population. Beneath each of these major Census categories hides a range of more subtle differences, subgroups, geographies, identities, 'write-in' labels, and so on, most of which will only be available for research within the next years. Furthermore, when these ten variables can be cross-tabulated, between themselves or with other socioeconomic variables, a myriad of categories will allow researchers to study not only the characteristics of London's strikingly diverse population, but more importantly, to analyse which outcomes actually matter for such population sub-groups, establish temporal trends, and isolate potential sources of discrimination and disadvantage.

Looking into the future towards the 2062 horizon, what it is almost certain is that the ethno-cultural diversity of London's population is set to increase. As a result of inter-mixing and cross-fertilization between cultures, languages, nationalities, religions, identities, geographical origins and so on, London's population will probably become one of the most heterogeneous cities in Europe. Furthermore, the unproblematic and clear-cut nature of the current Census categories will almost certainly look naïve in fifty years' time. Perhaps the failed prediction made in the 1920s by sociologist Max Webber, who stated that 'primordial phenomena' such as ethnicity and nationalism would decline in importance and eventually vanish as a result of modernisation, industrialisation and individualism, will finally become true in the London of 2062.

## Conclusion

How will London's population look in 2062? It will certainly look larger, older and more diverse than it is today. In this chapter, we have presented a range of evidence and broad trends, which together explain the dynamics that will in turn determine London´s future population. This exercise has indeed cast more questions than can be tackled in this short essay. Let this conclusion then be a wrap-up of intriguing questions and open-ended predictions that relate to other chapters in this book.

Will London be able to cope with the predicted increases in its population size? How and where will the jobs and houses required be created? The only option will be to increase population density in most parts of the city, a trend set since 1991. But when will the city reach its limit on infilling, retrofitting and recycling of brownfield sites? When will it be considered appropriate to demolish vast swaths of derelict Victorian housing stock to start building up a more compact and affordable city? This will surely bring political turmoil and expose London's stark inequalities between boroughs and neighbourhoods.

Despite physical constraints, the most important policy interventions will be those aimed at maintaining the socioeconomic and cultural factors that make London attractive to youths around the world today, as well as improving the environmental and economic constraints that push older people away. Only a vibrant, sustainable and liveable London will be able to keep its central role in the world city system beyond the mid-21st century.

London's population will look significantly different in 2062. It will be much older and much more ethnically diverse. The predominant ethnic group will be those with mixed heritage, for whom notions of identity according to a single national, religious or linguistic origin will seem remotely distant. As a result, it will also look extremely different to most of the rest of the UK, with important political consequences; perhaps a different migration regime could be applied to

London independent of the rest of the UK. This could for example open up the city to world migrants while implementing tele-surveillance measures to prevent residence elsewhere in the country. More likely, greater political powers will be secured for London´s regional government, who will finally manage to overhaul its deteriorating public infrastructure (transport, healthcare and schools) and prioritise building much higher residential buildings, over the decisions of individual local authorities and the interest of 'NIMBY' lobby groups. Against common perceptions today, such a dense and fluid city will be loved not only by city youngsters and the wealthy, but also by an aging majority that will not feel the need to run away from urban life when they have children; will have more than one place they call *home*; and almost certainly will never be able to afford full retirement.

# References

Atkinson S. 2006. Health inequalities in London: where are we now? Health in London - Looking back, looking forward. Available from: http://www.london.gov.uk/lhc/docs/publications/healthinlondon/2006/Section02.pdf. [Accessed 16 August 2013]

Baker A, Fitzpatrick J, Jacobson B. 2012. Capital Concerns: Comparing London's health challenges with England's largest cities. Available from: http://www.lho.org.uk/Download/Public/17872/1/Capital ConcernsRevised17.07.12.pdf. [Accessed 16 August 2013]

Bongaarts J, Feeney G. 1998. On the quantum and tempo of fertility. *Population and Development Review*. 24(2): 271–291. Available from: http://www.jstor.org/stable/10.2307/2807974. [Accessed 20 November 2012]

Cheshire J. 2012. Featured graphic: Lives on the line: mapping life expectancy along the London Tube network . *Environment and Planning*. (A)44: 1525–1528. Available from: http://www.envplan.com/abstract.cgi?id=a45341. [Accessed 15 November 2012]

Coleman DA, Dubuc S. 2010. The fertility of ethnic minorities in the UK, 1960s-2006. Population studies. 64: 19–41. Available from: http://dx.doi.org/10.1080/00324720903391201. [Accessed 15 November 2012]

Graunt J. Natural and Political Observation on the Bills of Mortality [Internet]. 1665. Society R, editor. Available from: http://scholar.google.co.uk/scholar?q=graunt+bills+of+mortality&btnG=&hl=en&as_sdt=0,5#3. [Accessed 20 November 2012]

Greater London Authority. 2011. Birth and Death Rates, Ward. Available from: http://data.london.gov.uk/datastore/package/birth-and-death-rates-ward. [Accessed 16 August 2013]

Greater London Authority. 2012. Births and Fertility Rates, Borough. Available from: http://data.london.gov.uk/datastore/package/births-and-fertility-rates-borough. [Accessed 16 August 2013]

Greater London Authority. 2012. Population Projections to 2041 for London Boroughs by single year of age and gender using the Strategic Housing and Land Availability Assessment (SHLAA) housing data. Available from: http://data.london.gov.uk/datastore/package/gla-population-projections-2012-round-shlaa-borough-sya. [Accessed 16 August 2013]

Greater London Authority Intelligence Unit. 2012 (October). Migration Indicators Intelligence Update 24-2012. Available from: http://data.london.gov.uk/datastorefiles/documents/update_24_2012.pdf. [Accessed 16 August 2013]

Mateos P. 2014. The international comparability of ethnicity classifications and its consequences for segregation Studies. In: Lloyd C, Shuttleworth I, Wong D, (eds.). *Social-Spatial Segregation: Concepts, Processes and Outcomes*. Bristol: Policy Press. (In press)

Office for National Statistics. 2011 (19 October). Life expectancy at birth and at age 65 by local areas in the United Kingdom, 2004-06 to 2008-10. Available from: http://www.ons.gov.uk/ons/publications/re-reference-tables.html?edition=tcm:77-223356. [Accessed 16 August 2013]

Office for National Statistics. 2012. 2011 Census, Key Statistics for Local Authorities in England and Wales. Available from: http://www.ons.gov.uk/ons/publications/re-reference-tables.html?newquery=*&newoffset=25&pageSize=25&edition=tcm:77-286262. [Accessed 16 August 2013]

Von Ahn M, Lupton R, Greenwood C, Wiggins D, Ahn M Von. 2010. Languages, ethnicity, and education in London. Available from: http://repec.ioe.ac.uk/REPEc/pdf/qsswp1012.pdf. [Accessed 16 August 2013]

# Making London, through other cities

## Jennifer Robinson

The future of any city is made and imagined in a world of cities – in a world with competitor cities, cities which could be adopted as models, or cities that might perform a future to be vigorously avoided. Cities are very often placed imaginatively alongside one another as we dream urban futures. Boris Johnson, Mayor of London in 2012, entrained the world of cities in his personal vision that London would be the 'best big city in the world' (Greater London Authority, 2011). The competitor cities of New York, Frankfurt and Tokyo loomed large as the detailed London plan articulated the potential for supporting strategic elements of London's economy. And the dynamism of Asia's urban skylines arguably stretched London's horizons upward as the previous Mayor responded to the sense of a fading urban image by changing planning regulations to allow tall buildings (McNeill, 2002) – the geopolitics of huge buildings is just one of the ways in which China is announcing its global ascendance. But Ken Livingstone's mayoralty also saw the creation of a co-operative network of large cities (the C40) committed to seeking creative solutions to the challenges of climate change (Gavron, 2007).

The prolific entwining of cities with each other will be a crucial determinant of London 2062. However, as these examples show, there are many different ways to approach such planetary entanglement. In the 2012 debates about London's airport capacity, competition and collaboration, real choices for the future were framed – should the city urgently extend airport capacity or lose traffic (and by implication economic growth) to other hub airports more able and willing to expand? Or perhaps could the intense connections amongst European cities be used to provoke a more interdependent solution, one where air traffic is shared across the continent and ground travel plays a bigger role? For geographer Doreen Massey London's prolific connections with many other parts of the world come together in the spaces of the city and interweave and overlap to shape the city's possible futures (Massey, 2007); they also expose the political choices we have for shaping London's future, and future co-existence on this planet. Her analysis resonated strongly with Mayor Livingstone's controversial negotiations with the Venezuelan government of Chavez for cheaper petrol prices for London buses – the affordable travel option which expanded considerably in his term. An unusual connection, perhaps, and one which was

**How to cite this book chapter:**
Robinson, J. 2013. Making London, through other cities. In: Bell, S and Paskins, J. (eds.) *Imagining the Future City: London 2062*. Pp. 23-26. London: Ubiquity Press. DOI: http://dx.doi.org/10.5334/bag.b

derided and terminated by Ken's conservative successor, but it speaks of the great diversity of arcs and flows which tie the fortunes of cities around the world together, from Caracas to Cairo, Nanjing to Nairobi.

Such is the interconnectedness of the world of cities that many urbanists now insist we think about urbanisation beyond the physical entity of the city – Henri Lefebvre's hypothesis of the 'complete urbanisation of society' asks us to think of our social world as profoundly urban, made through social relations which extend across the globe and result in urban forms which often stretch beyond any physical limits we can identify. As 'the urban' generates a planetary social form, it is the social and physical processes of urbanisation rather than discrete cities which should attract our attention (Lefebvre, 2003; Brenner & Schmid, 2012). According to many commentators, London's future will be made in a world of rapidly growing, extremely large and territorially extended cities – several cities now enumerate above 20 million, and London is likely to be a long way from being a 'big' city in 2062. It will probably be somewhat peripheral to the making of the planet's urban future.

Many other cities, then, are connected up in one way or another to the fortunes and lives of this city and its citizens. For London this worldliness is deeply embedded in the daily lives and dreams of its residents who have made their way to this city from so many different places, and whose livelihoods are closely imbricated with elsewhere. These interconnections are often seen as emblematic of the city as a whole: the city of Empire, the most multicultural city in the world, a centre of financial globalisation. The kinds of connections London forges with other places *now* matter for the future of the planet; they also make the future of this city. London 2062 will be made by political choices about how to engage with these wider connections, and how Londoners choose to cohabit in the midst of the many intersecting worlds of the city.

China Miéville, the novelist who writes London's future obliquely in his 'new weird' fantasy novels, comments that he is 'not interested in fantasy or SF as utopian blueprints, that's a disastrous idea. There's some kind of link in terms of alterity... If you think about the surrealists, the estrangement they were trying to create was a political act. There's some shared soup somewhere in my head from which these two things are ladling.' (Jordan, 2011) The estrangement which science fiction and fantasy novels effect as they take us from somewhere deeply familiar to a shocking alternative reality perform this politics of alterity – of imagining alternative, other Londons (Moylan, 2002). Miéville's award-winning novel (Pan Books, 2009) *The City and the City* takes us very quickly from a conventional detective scene to seek out (detect) the strange but achingly familiar truth of a city juxtaposed/superimposed on its other, whose buildings and inhabitants are routinely 'unseen' day by day as two coexistent and tightly imbricated realities are kept firmly apart.

This is a quintessentially London geography – the intricate mosaic of this city sees wealthy and poor, diverse ethnicities and many different language groups sharing streets, parks, festivals and often schools in an apparent outpouring of tolerance and multiculturalism. But this subtle proximity can also be seen to be managed through a measure of indifference and produces startlingly divergent outcomes in life chances, educational opportunities, housing quality and income – outcomes so differentiated that the parallel social worlds they invoke could come to resemble nothing less than the hard lines of segregation which appear so solid in some of the other cities which shadow this novel – Jerusalem, Johannesburg, Sarajevo. These divisions foreshadow South African novelist Lauren Beukes' equally astute but more clearly scifi *Zoo City* (Beukes, 2010) in which the segregation of today's Johannesburg reaches into a future defined by the abandonment of the city centre to second class criminalised citizens with companion animals assigned to soak up their dangerous psychological excesses.

Current trends, such as the corporate globalisation of London's physical spaces through eager regeneration, the displacement of working class communities and the decline of social housing stock are exacerbating the social inequalities of the city, creating a futuristic exclusion of the poor (Imrie et al., 2010) – a spatial inversion of the zoo city fantasy as the city centre is increasingly imagined as the preserve of the wealthy. 'Unseeing' (which might be figured as a kind of tolerance) and spatial segregation infect one another as political choices arguably describing a path towards a *noir* urban future.

The shock of Miéville's fantastic elaboration of the familiar practice of urban unseeing – reminiscent of Simmel's famous blasé attitude, or the discreet aversion of eyes familiar to anyone who catches the tube in London – juxtaposed with the extreme nihilism of Beukes' futuristic Johannesburg illuminates the political challenges of London 2062. The ways in which we choose to live together now, and the ways in which we practice the interconnections which make this city in the context of multiple elsewheres, will determine the future geography of this city, and shape the future of the world of cities.

## References

Beukes L. 2010. *Zoo City.* Oxford: Angry Robot

Brenner N. and Schmid C. 2012. Planetary Urbanisation. In Gandy M., editor. *Urban Constellations.* Berlin: Jovis. p. 10-13

Gavron N. 2007. Towards a Carbon Neutral London. In Burdett R. and Sudjic D., editors. *The Endless City.* London: Phaidon. p. 372-385

Greater London Authority. 2011. *The London Plan: Spatial Development Strategy for Greater London.* London: Greater London Authority

Imrie R., Lees L. and Raco M. 2010. *Regenerating London: Governance, Sustainability and Community in a Global City.* London: Routledge

Jordan J. 2011. A Life in Writing: China Miéville. *The Guardian.* 14 May. Available from; http://www.theguardian.com/books/2011/may/14/china-mieville-life-writing-genre. [Accessed 9 August 2013]

Lefebvre H. 2003. *The Urban Revolution.* Minneapolis: University of Minnesota Press

Massey D. 2007. *World City.* Cambridge: Polity Press

McNeill D. 2002. The mayor and the world city skyline: London's tall buildings debate. *International Planning Studies.* 7(4): 325-334

Moylan T. 2002. *Scraps of the Untainted Sky: Science Fiction, Utopia, Dystopia.* Boulder: Westview

# Hinterlands

## Brian Collins

Studies of the future of London almost always talk of its relationships to suburbs, green belt and the preservation of it, and the connectivity between London and other cities in the UK and elsewhere; these might be classified as the physical hinterlands of London. What will become increasingly important over the next five decades are the value propositions of other dimensions of 'hinterland': political power, economics and finance, trade and commerce, international diplomacy, resources and supply chains, and skills and education and the interactions between them.

The balance between the political power of the London Mayor and the Lord Mayor of the City of London, and the Prime Minister of the UK National Government is already showing signs of change. As the social and economic activity of London becomes a more important share of overall national activity, recognising the impact of this change will be critical. This impact will be on a range of factors, all of which interact with each other. The pragmatic approach of reacting to pressure for change is one route forward (that seems to have been taken up to now); but for leaders to be informed by analysis of the sensitivities of possible futures to their political decisions or indecision would seem to be a fruitful avenue for research.

The importance of economic and financial activity in London is obvious; what is less obvious is its sensitivity to activities elsewhere on the planet. Whilst the financial markets service the domestic markets, London as a global hub services a very significant fraction of global markets in a range of financial instruments. This came from London's past dominance of finance, particularly in insurance and shipping sectors. Is it reasonable to expect this situation to continue? And what are the conditions for this assumption being valid? As other cities develop, they will grow a capability to compete in this hinterland of financial and economic activity, especially as the centre of gravity of trade moves to Asia. What skills and experience does London have that can be sustainably maintained to offset this growth elsewhere, or should London plan for a reorientation of its financial markets? History shows that cities that depended on one major industry without planning for a more efficient future or diversification do not do well when more efficient competitor cities enter the marketplace. London has managed this transition in the past, but now needs to continue to do so, and at an accelerated pace in the near future; it cannot afford sclerotic public administration to be a barrier to progress and innovation

**How to cite this book chapter:**
Collins, B. 2013. Hinterlands. In: Bell, S and Paskins, J. (eds.) *Imagining the Future City: London 2062.* Pp. 27-29. London: Ubiquity Press. DOI: http://dx.doi.org/10.5334/bag.c

The brand of London is projected worldwide, and its culture, arts and heritage are known the world over; this benefits the tourism industry significantly, which in turn is supported by good but diminishing air transport links. But as other nations develop their tourist industries based on their own heritage and cultural assets, they will capture a greater percentage of a more global tourism business. This will not be helped by London allowing its air transport links to decline. This is one stark example of where London is a node of a global network, in this case air transport, where maintaining the ability to exploit the network is critical to sustainability. Connectivity throughout the global aviation hinterland is crucial to survival, and this equally applies to commerce and trade where most goods leave and arrive in our ports by ship. The understanding by London of its dependence on other airports and ports for a wide range of goods and services is also critical. Modelling possible futures, based on good data from across the world, will allow London governance to be executed in such a way that the resilience of London to shocks elsewhere is understood, and its ability to exploit opportunities elsewhere is sharpened.

Crucial to success for all these factors is a body of well-educated and skilled people who are motivated to make London one of the best cities in the world to live, work and raise families. So the 'hinterland' of schools, hospitals, universities, places of work and leisure are part of the value proposition that a sustainable development agenda for London must contain. Such an agenda is not new, but sustaining it over five decades is a political challenge which has yet to be successfully met.

Almost all of these issues are not new; what may be more novel is consideration of them all at once and in a 'joined-up' way, and to then maintain continuity of policy and of execution to 2062. This will demand sustained leadership and vision not seen in London (or, it might be argued, in the UK) since the end of the Second World War. A master plan for London has been created many times; were such a plan to be created again it should include all the factors of hinterland described above, because of globalisation, the situation is more complicated than before. But without such a plan London will continue to adopt pragmatic short-term solutions that will be less than optimum for London and its hinterland and hence cause progressive 'deep water' to open up between London and its competitor cities. A master plan does not have to be executed centrally, nor is it set in concrete. It can be executed locally in a way that is compliant with the overall plan, and it can be reviewed at a sensible periodicity; say every five to ten years.

A master plan provides a common aim point for all activities in London; then all our futures as Londoners might be provided through a process that is more like a journey in daylight with a map than an expedition at night with a torch.

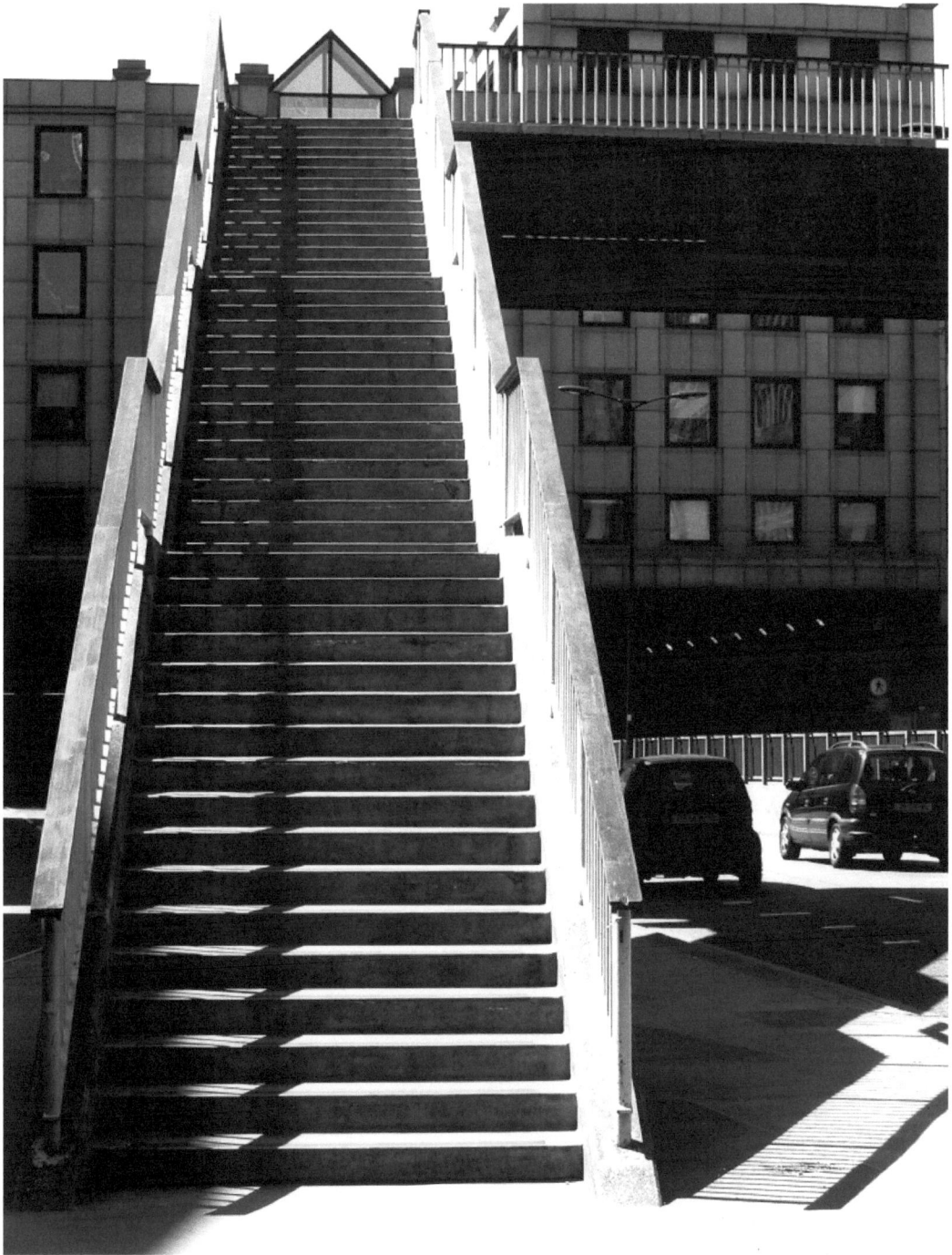

# Smart London

## Michael Batty, Ed Manley, Richard Milton and Jon Reades

### Technology, information and cities

Suddenly, after sixty years of dramatic advances in digital computing where computers are now being embedded and networked in every imaginable artefact including ourselves, cities for the first time are becoming automated. Cities in fact are turning into computers with enormous and unprecedented effects on how we behave and function within them. At every stage of the computer revolution since the 1940s, we have been reminded that the computer is a universal machine. The father of modern computing Alan Turing in 1948 stated that "A man provided with paper, pencil, and rubber, and subject to strict discipline, is in effect a universal machine" (Turing, 1948). This idea, reduced to its bare essentials and embodied in silicon, set the world on its way to our current age where computers are more numerous by far than their users.

The size and scale of cities is intrinsically linked to their technology. Cities were barely able to grow beyond one million in size before mechanical technologies were invented at the beginning of the industrial revolution, and only then was it possible to ever envisage more than a handful of cities of this size. In fact at any one time there was rarely more than one such city. Rome for example, grew to about this level but it was sustained by an enormous empire which at first was held quite tightly together by highly disciplined communications bolstered by the army. After Rome, Nanjing and Beijing reached similar sizes in the 6–7th centuries while London only reached one million in 1800, before industrial technologies began to revolutionise communications.

Once mechanical technologies became pervasive, then cities began to grow, first through mass transit based on rail and streetcar which opened up and spread their hinterlands and then through the automobile. Cities for the first time reached ten million in the mid-20th century, New York being the first, and they also grew in the vertical direction, again due to new technologies based on the steel frame and the elevator. The great conundrum of course is what will electrical technologies do to the size and shape of cities? For over 100 years, the telephone has been key to linking places together but now with email and various social media, web-based interactions are becoming near

**How to cite this book chapter:**
Batty, M, Manley, E, Milton, R and Reades, J. 2013. Smart London. In: Bell, S and Paskins, J. (eds.) *Imagining the Future City: London 2062*. Pp. 31-40. London: Ubiquity Press. DOI: http://dx.doi.org/10.5334/bag.d

universal, offering prospects of remaining social and economically connected anywhere, everywhere and at any time but always in the context of large scale physical disconnection. Although we have little clue as yet of what this will mean, it must have an important impact in the next fifty years on the size and shape of our cities and how we organise ourselves spatially.

London of course is a fulcrum in these changes. For 200 years or more, it has been a global city and it is rapidly becoming a city-state in its own right, spreading far beyond its physical borders and in some sense, being the core of a highly integrated urban system in the UK, and beyond. Increasingly we need to think of such a city as a city of flows manifest at many scales from the local to the global, envisaging its future as one in which networks will dominate locations, flows being the new currency in which we should think of all cities as opposed to locations or hubs. Of course we still need to picture cities as sets of structured locations but it is ever more important to think of cities as constellations of flows, not unlike the definitions of the near future, of cyberspace, by Gibson (Gibson, 1884) when he says that (cyberspace) is: "A graphic representation of data abstracted from the banks of every computer in the human system. Unthinkable complexity. Lines of light ranged in the nonspace of the mind, clusters and constellations of data. Like city lights, receding." This is as good a definition of what cities are becoming as any and the kind of science that is needed to progress these ideas is on its way (Batty, 2013).

These changes which will be writ large in a city such as London within the next fifty years, indicate a very different relationship between such large world cities and their hinterlands than anything in the past. London effectively will act – indeed *is* acting as the global portal for every other city in the UK and perhaps for other cities in its extended global reach. The idea that other big cities should try and compete with London for this reach is problematic, in that this suggests that London (and every other city of this size) is a sensitively-engineered organism whose survival and prosperity depends on a much wider integrated system than one would ever imagine by simply looking at its physical size. In this sense, there is an intrinsic logic to the way globalisation has evolved and it is essential to account for this in any strategy that pits one city against another.

## Embedding computers in cities

In large cities, of which London is the contemporary archetype, the focus is as much on what is happening inside the city with respect to the use of technologies as it is on enabling large cities to become global hubs in the world economy. Computers are being fast embedded into cities to restructure their functioning: to automate urban functions that traditionally were operated largely manually and mechanistically. Instrumentation which enables vehicles and structures to communicate with one another is now becoming widespread. This has happened rather quickly as computers have been miniaturised to the point where they can be embedded into any object. How people communicate and transport themselves, their goods, their information, their ideas, is of course the most obvious of contexts where this has occurred most rapidly. Transport systems where people and vehicles move in regular and routine fashion are now controlled automatically using diverse computer systems whose data is provided from real-time sensing of position and time. We now know where every vehicle and indeed every traveller in the city is, in terms of many well-defined public transport systems, and it is only a matter of time before this will be true for most physical movements. This will potentially create a revolution in how we travel and how we can use information about the system in general to make instant decisions about how to improve our travel experience. Within the next two or three decades, there will be a revolution in the home where all devices that enable us to control the energy we use will be automated through smart metering, which is happening even faster in office and industrial environments. In the short term, as James Martin (Martin, 1978) said more than a generation ago, society and its cities are becoming 'wired' with all the implications for how we might behave in such environments now at the forefront.

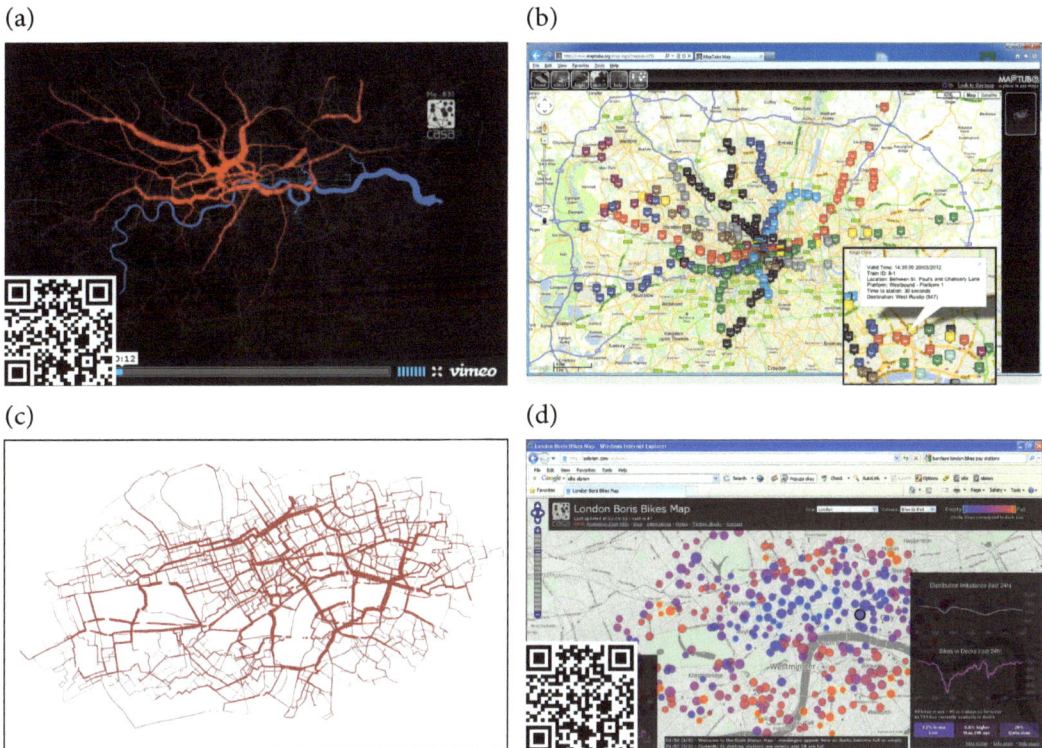

**Figure 1:** Visualising Smart London 2012.

(a) Public transport flows over one week from Oyster card data. The video by Jon Reades is available at http://dx.doi.org/10.5334/bag.d.1 (or by scanning the QR code in the bottom-left corner); (b) Tube trains in real time (c) Bike flows from the Barclays public bikes scheme. The video of the bike flows from Martin Austwick is available at http://dx.doi.org/10.5334/bag.d.2 (or by scanning the QR code in the bottom-left corner); (d) Bikes in docks at any cross section in time from Ollie O'Brien where you can access up to 100 schemes as well as London and play back animations of the demand and supply of bikes over the last 24 hours http://www.youtube.com/watch?v=NSO4EScYcj8.

There are many examples in present-day London which illustrate how the city is becoming 'smart'. Smart card data from the stored value system called Oyster Card now records all data on where travellers enter the public transport system and leave it. We have months of data on the movements of some six million travellers per day who use the tube and overground rail and we can routinely animate these flows. We also have real-time data from the interfaces that record and communicate information on public transport systems such as bus and rail on the position and delay of any particular vehicle, thus providing us with an instant picture of the supply of public transport at any place and time, as we show in Figure 1(a) for the demand on the tube.

In short, we have data on demand from the Oyster Card and on supply from these interfaces and in principle we should be able to connect these up. In practice, there are enormous problems in doing this because we do not have detailed data on how travellers navigate and use these systems and how long it takes passengers to get on trains, and of course we do not have the actual data relating to individual passengers on trains but in principle we can reconstruct this. Very likely within the decade, we will have much better ways of effecting such integration but at present this is a major obstacle to making these systems smarter. We show the position of the tube trains at a snapshot of time in Figure 1(b) and the challenge faced in linking this data to that shown in Figure 1(a).

The data now extends to public buses and it is a straightforward matter to integrate this with other road traffic volumes. The public bikes scheme in London has been online since its inception in 2010, again illustrated in Figures 1(c) and (d). In fifty years, it is certain that all of this data will

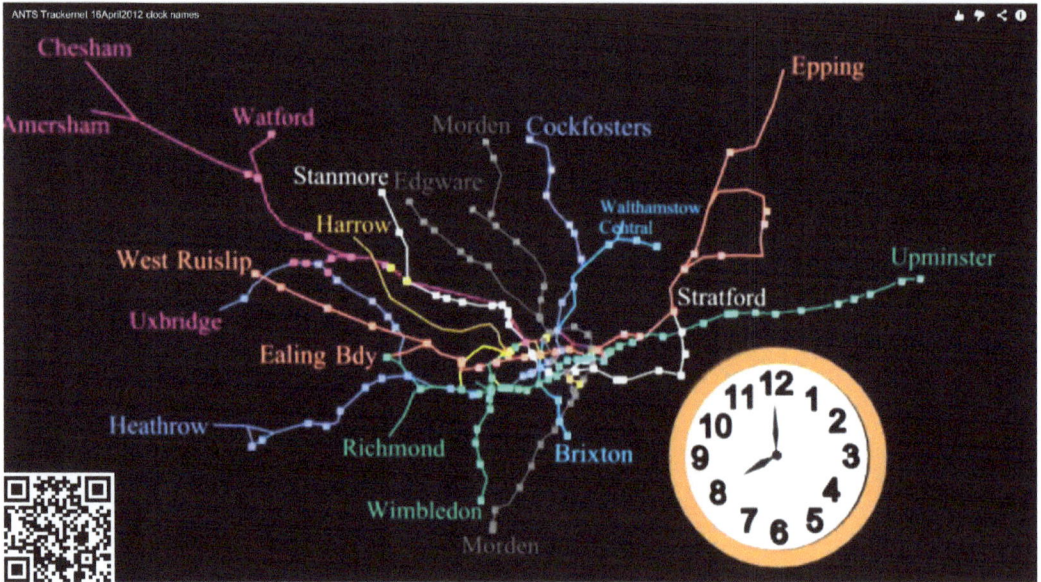

**Figure 2:** The Flow of Tubes (and Buses) in Real Time.
The animation of the tube trains over a 24 hour period, speeded up of course with some commentary on the data is available at http://dx.doi.org/10.5334/bag.d.3 (or by scanning the QR code in the bottom-left corner).

be integrated and will provide an instant picture of traffic conditions throughout the city. Many other flow systems will be integrated with this such as where pedestrians walk, the locations of hotspots focussing on accidents, crime, access to services, and a host of other data. Integration will remain a problem because these data sets are usually collected either remotely or, if they are related to population attributes, these attributes cannot easily be integrated across different systems. Moreover the technologies themselves are continually undergoing change as are the systems used to collect such data and there is no guarantee that systems such as the ones we now have will continue in the same way into the future. This is a problem of the city becoming too smart for its own sake as new systems take over from those in place, transforming and destroying the data volumes and the experiences that have been gathered so far. These will pose major challenges.

To give an illustration of what will become routine in large cities like London within the next fifty years, if you click on Figure 1(a), you will see the demand for public transport on the tube over a typical week where the movie clearly identifies the weekdays in comparison with the weekends and where the morning, evening and late evening peaks are clearly visible on weekdays. Note however that these data sets show the great heterogeneity of the data from minute to minute as no flow from one day to the next is the same. The same is true of the bikes data in Figure 1(d) where, if you click on the link, the bikes data is animated for a typical 24-hour period just previous to the time when you log onto this web link. In fact we can show some of this data in real time and if you click on Figure 2 you will be taken to a site which shows the animation of this data and thence to a site that shows the movement of tube trains and buses in real time: the tube trains are very slow as they are moving in real time at the scale of the web page and of course they are much bigger on this scale than the map itself simply so we can show them.

This shows London Underground trains for 16 April 2012 between 8am and 8pm using data taken from the TfL Trackernet API. A disruption can be seen between 08:30 and 09:00 when everything stops and then the tubes all re-position themselves. This was caused by an outage in the Trackernet API which feeds us the data. Other anomalies can be seen where tubes move off of the network and then re-join. This occurs where a train is re-positioned part way through its route as

(a)

(b)

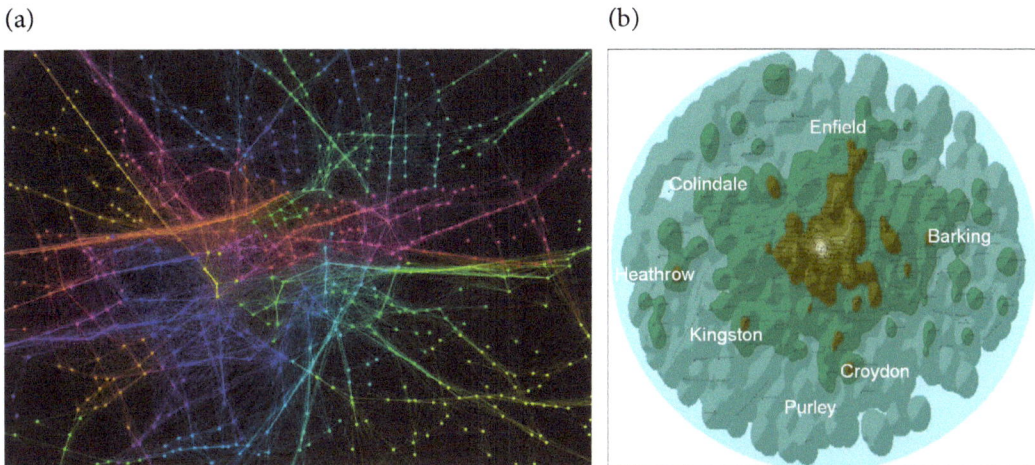

**Figure 3:** Detecting Location and Network Clusters in Digital Media.

(a) MiniCab routes in central London, where homogenous regions and routes naturally emerge as groups within the dataset, particularly identifying important thoroughfares such as Euston Road, Park Lane and Embankment and areas such as Soho, Shoreditch and Kings Cross; (b) Tweets plotted over 24 hours on a typical weekday in mid 2010 in Inner London, illustrating the tendency for higher densities to define entertainment clusters; data collector software by Steven Gray, and visualization by Fabian Neuhaus (CASA, UCL).

Some animation of tweet data with respect to key locations in the 2012 Jubilee celebrations by Ed Manley are shown at http://www.youtube.com/watch?v=M15SqwlSt1c and http://www.youtube.com/watch?v=2pYP5fiy8Rg. There is a write up in the Guardian newspaper at http://www.theguardian.com/news/datablog/2012/jun/11/queen-s-diamond-jubilee-twitter-tweeting.

it disappears from the system for a period of time and then re-appears somewhere else. Although this could have been removed from the animation, it is useful to see where it is happening and has been used to diagnose problems with our tracking system.

The data we get from TfL does not identify the line that a tube is on, so we rely on identification from the location and destination codes. Occasionally, tubes can be misidentified, but the system is adaptive in the sense that it has a limited ability to detect this from the data and modify its destination code table for the next time.

If you want to see the real time web site, please contact Richard Milton (Richard.milton@ucl.ac.uk) who is the author of this but currently the site is not suitable for direct access. In time, it will be and this is something that will be available within the next decade for routine viewing.

There are many new data sets too emerging from all kinds of physical networks transmitting materials and people as well as social media that implies various kinds of social network. In Figure 3(a), we show the communities which can be defined for various kinds of business and related travel from real-time digital data on an up-market limousine (minicab) system in Greater London and this implies how we might use this kind of network data to examine the multiplicity of movement patterns in the city and how these determine locational hotspots. In contrast, in Figure 3(b) we provide a picture of tweets in London over a typical weekday which can be geo-located, giving some sense of the density of tweets with obvious comparisons with related activities and land use clusters. In the future city, in future London, we will see many of these patterns being generated in real time and integrated with one another to give us a new understanding of the way the city functions. In Figure 4, we show the pattern of tweets by language over the summer of 2012 in London, and if you drill down on this, you can zoom into different areas and examine the locations of these languages in detail, comparing your knowledge of what goes on in that part of London with this data. This Twitter data is as good as it gets at present, and language is fairly uncontroversial. It implies location and probably residential

**Figure 4:** The Twitter Languages of London.
The language of tweets sent from the London area over the summer of 2012. Of 3.3 million tweets, 92.5 per cent are, not surprisingly, in English. The biggest tweeting tongues after that are Spanish (grey), French (red), Turkish (dark blue), Arabic (green), Portuguese (purple), German (orange), Italian (yellow), Malay (turquoise) and Russian (pink). If you drill down you can zoom into the web page where these tweets are easier to locate and identify: http://spatial.ly/2012/10/londons-twitter-languages/ and at http://twitter.mappinglondon.co.uk/.

location but this kind of social media is presently quite problematic in what it can say about the workings of the city. But it is early days and in time, much more will be possible from this kind of data.

This kind of data that is both a by-product of the smart city and also an essential determinant in making it smarter is changing our perception of cities and producing a much more immediate sense of how a city functions. Big data like this gives us the opportunity of looking at short time scales and this is changing the focus to questions of the short term such as disruptions and disjunctions, to identify differences and hotspots which pertain to good and bad functioning in the city. To an extent, we are currently experiencing rather unusual new forms of data and information which in turn pertain to new kinds of functioning of our physical infrastructures. The great challenge of the next fifty years will be putting in place systems to best exploit these new ideas and to use them to further better the quality of life in cities. Technology is clearly central to the city and the biggest question is: What will places like London look like in fifty years' time? We will close with some speculations.

## How smart will London be in 2062?

If you look back more than half a century to London in the post-war years, technology in the city was very different. In 1946, there were hardly any cars but at least it was widely accepted that cars would become affordable for the majority of the population within a couple of generations. The

American experience in the 1920s showed that this was possible. However a more revealing statistic was that there were but a handful of computers. Indeed the ten Colossus computers built by post office engineers at Dollis Hill in Northwest London for the code-cracking activities at Bletchley were just being destroyed by a War Office mandate. It took until the mid-1980s for computers to become individually available through the personal computer, and fifty years ago in the early 1960s there was little sense that within a generation the city and the world would be well on the way to being underpinned by widespread computation and the information that this would bring. Back then, the impact of the car on the city was only just beginning to be felt, and even now when computers are everywhere, we are only beginning to grapple with the transformations in physical structures that these will bring in the next fifty years.

The great challenge of course in envisaging London in fifty years' time is in knowing what the balance will be between hardware and software. History suggests that the focus will be entirely on information, that hardware will be writ large but passive and pervasive, but quite what this will mean is hard to grasp. In terms of how it all looks, maybe we will all be wearing Google Glass and sitting in Google Smart Cars. The fact that these exist is an issue but the extent to which they will be all-pervasive is debatable. Reactions against social networks and technologies that are intrusive will be a strong feature of this future and it may well be that some of the science fiction-like technologies such as driverless cars on automated driveways will simply be too difficult to implement. Our experience of building integrated computer and information systems so far has not been good and the difficulties of merging the most obvious data sets such as demand for and supply of travel, as we have indicated above, are legion.

Yet the world we have sketched at present with respect to sensing and big data will massively increase. Imagine then a world, which will not be too long in coming, where all data of this kind can be integrated and accessed and then visualised at any moment, thus providing an immediate picture of how the city functions. Add to this data on energy use in every building, again in real time, and a picture of how sustainable the city is in these terms quickly emerges. Of course, the grand challenge is to make more invisible data on how people relate in terms of their financial transactions accessible, so that we have a much better view of how people and other agencies are engaged in various economic markets such as housing and land. These markets continue to fail to provide the right levels of supply and we urgently need to figure out how capital flows in cities so that we have some sense of where key problems lie. To an extent, such data is becoming available through financial transactions. The Senseable Cities Lab at MIT has visualised credit card transactions over short periods in Spanish cities and there is a well-known web resource for tracking dollar bills in the United States (Batty, 2012). Information which is beginning to substitute as well as complement material flows is much harder to gauge and measure but cities and society now function through such flows. In the next fifty years, we will see enormous strides in measuring information and being able to see how it relates to the space-time functioning of cities. Alongside more visible flows such as transport, this will be the great frontier in enabling our cities to react more effectively to change.

The prospect that there will be much more automation in spatial markets such as housing and really up-to-date information about where populations locate and migrate suggests a future which is much more liquid and volatile than the present, and it is possible that our future concern will be with trying to design into such systems feedback effects for damping such change. Whatever systems emerge, it is clear that we will have a much better picture of real-time populations as they change in space and time over the day, the week, the month and so on and over local to global spaces as people travel and migrate.

Of course in fifty years' time, we will be in a position where much of the data that is now being collected automatically will inform our longer term quest in cities, which involves figuring out traditional problems of where the city is headed. Already we have rudimentary models of how cities function at the aggregate level and recently we have been involved in an integrated assessment

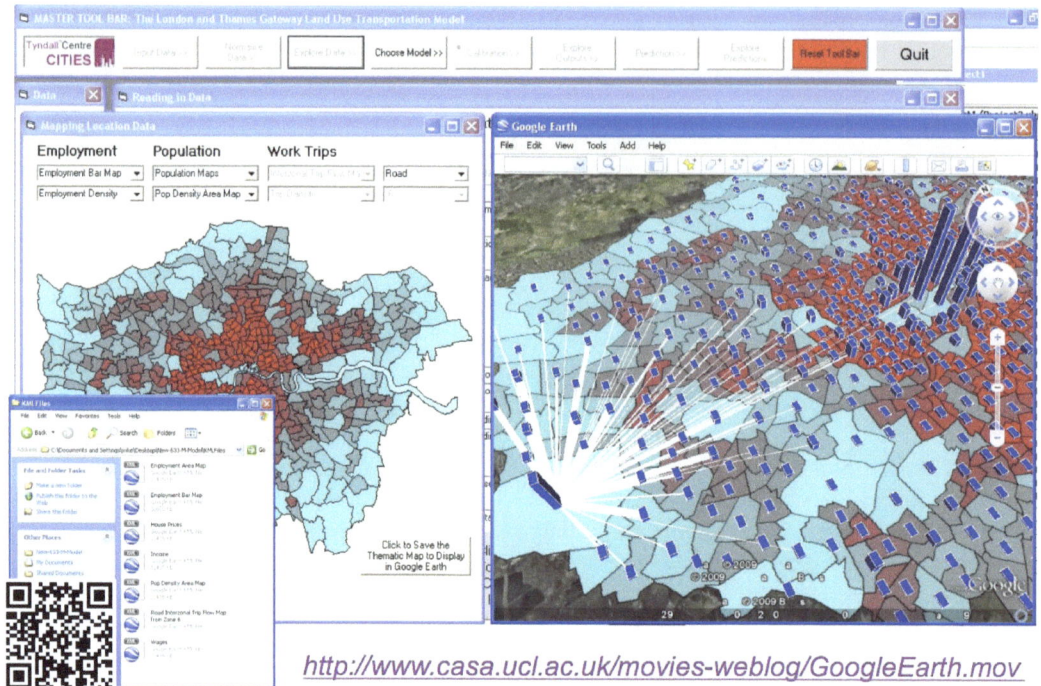

http://www.casa.ucl.ac.uk/movies-weblog/GoogleEarth.mov

**Figure 5:** Modelling London 2062 in Real Time?
Our urban model of Greater London can predict locations of future population quickly and graphically. Imagine this kind of forecasting in real time, on-the-fly, continually replenished with new data on a daily basis and accessible anywhere. The only question is whether we want such tools, not whether they are possible. The movie is available at http://dx.doi. org/10.5334/bag.d.4 (or by scanning the QR code in the bottom-left corner).

of the impact of climate change in the London region where the key issue for the next fifty to one hundred years is sea level rise, hence flooding. We have various models that make predictions in small areas of the city where future population and employment will locate (United States Currency Tracking Project, n.d). Fifty years is a long time and no one can pretend to know how we will react in adapting to and mitigating such climate change, but it is now widely accepted that predictive modelling of the kind that has been developed for cities since the 1950s can be used to inform the debate. In London 2062, we imagine that the models we are currently using will be routinely accessible and operational, that we will be examining long-term change on a routine basis, using models of the kind we imply in Figure 5 to inform us continually of how the city is responding and changing. If we think of future London as a moving target, then our models can be updated daily in terms of their data, now from routine sensing, and thence operated in control-room manner in the same way that London and other big cities now manage their traffic systems.

Although what is happening to make the city smarter is largely occurring on a routine basis through new forms of sensing, the next years will be ones where much of this data is integrated. New sources of remotely sensed data will provide daily, or even instant, visualisations of the city and our more abstract scientific models will be integrated in such a way that they become accessible to many more groups of stakeholders. Imagine that each day the data for these models are updated and conditional 'what-if?' scenarios forecast all the time, for short periods of days to speculations about what might happen over the next fifty years. Not only data but also predictions of the many-term futures that the city will face will become available directly on whatever devices we then have to view such information. And as a consequence of this, many, many opportunities will exist for moulding the future city in online forums through various sorts of participation.

Whether or not we take these up depends on much wider considerations of how we will govern ourselves in an age when whatever we do will be informed by information technologies.

The London of 2062, in terms of how we understand it, will be informed by all these technologies. Given the rate of change in computing since 1962, we cannot begin to imagine what this might actually mean. In fact, although the idea of running models continually and having all this data available for everyone will certainly be feasible, there are still enormous challenges in figuring how all this will play out. Most of all, perhaps, privacy and confidentiality are top of the agenda but we also know that by the year 2062, the problems we will encounter will be very different. Migration is likely to be a major concern while population growth will largely have ended, at least in aggregate terms. We will have passed what Ray Kurzweil calls the singularity (Batty, 2010; Kurzweil, 2006), when medical advances will be integral to the very way we behave and when longevity will be central to the way we figure out how we will live and work. Industrial society and the industrial city will be long gone, but we have little idea of what space and place will then mean. Our models and tools will of course adapt to all of this but what is clear and what we have learned from the last century is that we now know that we will never know the future. Although we might speculate that technologies will be central to this future, London in 2062 will be a very different place functionally, if not physically, from what it is today.

## References

Batty M. 2010. Integrated Models and Grand Challenges. ArcNews. Winter 2010/2011(4): 32. Available from: http://www.esri.com/news/arcnews/winter1011articles/integrated-models.html. [Accessed 2013 August 6]

Batty M. 2012 (28 April). Financial Footprints: Transactions in Real Time. Available from: http://www.complexcity.info/2012/04/28/financial-footprints-transactions-in-real-time/. [Accessed 6 August 2013]

Batty M. 2013. *The New Science of Cities.* Cambridge, MA: The MIT Press

Gibson W. 1984. *Neuromancer.* New York: Ace Books

Kurzweil R. 2006. *The Singularity is Near: When Humans Transcend Biology.* Penguin Group (USA) Incorporated

Martin J. 1978. *The Wired Society: A Challenge for Tomorrow.* Englewood Cliffs, NJ: Prentice-Hall

Turing AM. 1948. *Intelligent Machinery.* Teddington, UK: the National Physical Laboratory

United States Currency Tracking Project. N.d. Where's George? Available from: http://www.wheresgeorge.com/. [Accessed 6 August 2013]

# Flux and flow

## Christine Hawley

The apocalyptic fate of the subsurface dwellers in EM Forster's short story 'The Machine Stops' (Forster, 1909) and Thomas More's 'Utopia' (More, 2010) may portray scenarios of unparalleled difference, but both the fifteenth century and the nineteenth century writers were fired by an imagination about the future. The Forster cameo is one of foreboding, describing a catastrophe of unimaginable proportions where one might find parallels with the environmental cynics who portray scenarios of extreme flooding, extreme heat and cold that would leave communities decimated (Forster, 1909). By contrast, More's vision outlines a societal model of the future that addresses how to combat the social inequalities. His essay is prescient in that it understands profound social tensions that prove to be timeless. The answers he proposes take place on the island of Utopia where philosophically driven structures of governance and moral codes help to determine an ideal world (More, 2010).

Even if one's view is not as idealistic as More or as apocalyptic as Forster, it would be reasonable to assume that some measure of each is more likely than either in entirety, and that in this context we could ask the questions about the city of London and what it might be like, fifty years hence. What is undeniable is that the scenario planners must adopt a measured response to climate change, but designers must also think about the technological and socioeconomic adjustments that would influence behavioural patterns.

At this point in time it is difficult to establish data that can accurately pinpoint possible scenarios; environmental science offers, if not conflicting, then highly polarized evidence. The economic turbulence in the last century has failed to deliver models that facilitate better forecasting. The social and cultural shifts currently evident on a global scale may produce a new political order that none has anticipated. It is likely that in fifty years' time, much of the building fabric of the city will remain; however whether there will be sacrificial policies that benefit the historic or politically important, one can only speculate.

Scenario planning relies on considering a wide range of possibilities, some of it predicated on existing data which could be manipulated in a number of ways offering no one simple solution. The understanding and control of climate related issues are an inexact science, yet there will be both political and economic incentives to ensure that structures are insulated and stabilized. Much

**How to cite this book chapter:**
Hawley, C. 2013. Flux and flow. In: Bell, S and Paskins, J. (eds.) *Imagining the Future City: London 2062.*
    Pp. 41-44. London: Ubiquity Press. DOI: http://dx.doi.org/10.5334/bag.e

of the attention in the next fifty years will focus on existing building stock and infrastructure and how this might be safeguarded, but this will be selective; it would be wrong to assume that all structures are able to perform in an energy efficient manner. There will be those buildings that for historic and cultural reasons that will remain unaltered. The opportunities for designing and building new will be limited; there will be a requirement to utilize recycled materials and possibly the use of new composites that would create buildings with minimum hydrocarbon impact. The understanding of 'new' may become a completely altered concept.

Digital technology will have its impact on all aspects of life, from healthcare to education; from transportation to communication. The advances made, many already in trial, will have extraordinary impact and for many, will create unforeseen consequences. If we speculate that these changes could be serendipitous, how would this affect London and how would the architectural community prepare and respond? The current pattern of city use will be unrecognizable. Commuter traffic will significantly decline as transportation costs rise; commercial space will be redesignated, and there is the possibility that these premises could be re-colonised as dwellings. At the moment the dominance of the digital highway lies in informatics, yet it will only be a short time before all forms of fabrication are possible to control remotely. Therefore, let us speculate that the segregation of management, dwellings and production no longer exists. There are of course certain pragmatic issues, but the delineation of function geographically as we know it will no longer be relevant. If this happens, it will change the design agenda. Designers will become part of larger, integrated teams across the major disciplines, and the responsibilities that each will take may become unrecognizable.

If we consider some of the impact that environmental issues will have on London, the results will not only be architecturally pragmatic, but could also fundamentally alter the way we use the city. In fifty years, the concept of movement and why we need to move will have already been challenged; passenger travel could be airborne in vessels powered by inert gas. Notwithstanding the need to change the demand on fossil fuels for transportation, there is an argument to become much more radical in our thinking and not assume that road and rail travel will still be the backbone of our movement systems. A rise in water level might suggest the reopening of currently defunct waterways (both under and over-ground), and the increased use of existing water routes. There is a persuasive argument that indicates that the movement of heavy loads is done far more energy efficiently on water than on surfaces that create greater friction. It would not only be vessels that would change their energy source but the way we create energy would also be the function of London's waterways. Whatever advances are made in the science of harvesting alternative energy, the water courses of London will have a significant role harnessing energy in turbines through currents and tidal flow. The use of water-powered machinery at a micro level has been understood for centuries, and therefore the increase of micro hydropower would be feasible, even in domestic environments. The water tributaries will become critical pathways through the city, but these will also be used to develop new aquatic landscapes.

The supremacy of London parkland may give way to both public and private aquascapes that could both be used at a micro level for filtration, and as a new component in urban farming. Strategic and local groups will be implementing policies that encourage the use of open space for agricultural or aquatic cultivation. Data suggests that current farming practices will be economically unsustainable within fifty years, and it is for these and associated environmental reasons that farming within the city may become a reality. The image of the urban allotment has been a familiar picture since the interwar years; however the urban smallholding that farms meat and dairy produce has not yet become an established part of life. Within some new housing developments, it is arguable that the integration of a mixed farming unit is not implausible. During the winter months the animals would be kept indoors, and during the summer months they could get access to open rotational grazing in the existing parklands. The (treated) waste byproducts of such a system could be used to supplement the fertilizing requirements of conventional crops. To assume that all organic production would take place on horizontal surfaces is a limiting view; both new and existing buildings could be used to grow plants vertically using hydroponic systems. The production of organic material will become more incorporated into the fabric of the city and not just the agricultural plants we are currently familiar with, but autotrophic plants such as algae which will not only be used to create energy but also be used as a nutritional alternative to conventionally produced food.

The socioeconomic balance in fifty years' time has a number of unpredictable factors – the changing demographic and its cost to society, the reconfiguring of the workplace through digital advances and the nature of the manufacturing industries. The structures of communities are set to change. These structural shifts will have a fundamental effect on the ownership of property in the capital and perhaps, the way in which we conceptualise 'home'.

Prior to the first and second world wars, the housing stock barely moved in value; the destruction of the housing stock during the wars dramatically devalued the market. As such a high proportion of housing stock was rented there were particular pressures on landlords, who were often holding negative equity. Pre and inter-war tenancies were fixed and properties, if they were sold at all, were done so at a fraction of the cost of living index. What is interesting to reflect on – and perhaps a lesson for the future – is that at that time absolutely no one predicted the repeal of the 'Rent Act' and the enormous changes that were a result of this legislative decision: the social changes and modification of taxation law influenced a pre-capitalised market that precipitated significant rise in owner occupation. The model of housing may therefore radically change once again; the economic context will alter to challenge the traditional concept of owner occupation. If one were to link this together with mutable demographics and the rise of a dependent community, there would then be a requirement to design for extended family communities in central London.

There is a strange paradox, in that the enormous advances in technology and the sweeping fluxes of the economy could create patterns of living within the city that are akin to a historic model.

The physical vision of Forster predicts a society that must live beneath the surface of the earth as a result of the abuse of nature, and is finally annihilated when the technology that supports them fails [1]. The morality of this tale lies in the danger and power of human greed and the unconsidered consequences of technological advance. More's vision relies, more optimistically, on society's ability to organise communities of greater equality, where harmony comes through understanding that a multifaceted society benefits from interdependence rather than independence [2]. Perhaps we should reflect on the central themes of both and consider not only the physicality of our cities, but how technology, nature and governance will shape London in 2062.

## References

Forster EM. 1909. *The Machine Stops.* UK: Oxford and Cambridge Review
More T. 2010. *Utopia.* 3rd ed. WW Norton & Co

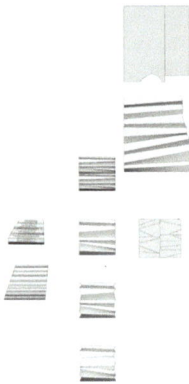

# Planetary pressures

## Jean Venables

The Future is not what it used to be!

Despite all our computer models of future climate scenarios and predicted effects, they are only possible scenarios. We live at a time of instability – financial, economic, social, and climate – which is making unprecedented and increasing demands on our planet. But instability can be a great time to trigger journeys in new directions, so what might the journey over the next fifty years look like?

The demands on our planet have been measured by eco-footprinting. If everyone on the planet were to consume and waste resources at the same level as UK citizens, we would need three and a half planets. Last time I looked, we just had one. So; whilst our inventiveness, creativity and technology will help, peoples' expectations and behaviours *must* change to adapt to one planet living, to a low carbon economy and low carbon living, to the climate change we have already, and to the climate change to come. How we respond will be crucial to survival; not of the planet, but of the human race. The effectiveness of our mitigation of the cause of climate change will govern how much adaptation will be necessary to adjust to rising temperatures, sea levels and wind speeds.

One important recent action has perhaps set a path for how to plan for and implement adaptation strategies in the face of uncertainty. The Thames Estuary 2100 Project sought to identify how to manage the flood risks to London and the Thames Estuary over the next 100 years. The outputs plotted the actions to take against possible sea level rises, and related the trigger times for starting intervention measures to the trajectory of observed sea level rise in coming years. This approach can be replicated not only in other major estuaries of the world, but for other types of infrastructure that will need to be upgraded to cope with the effects of climate change. The TE2100 Project outcomes also emphasised the vital need to maintain our current flood defences, which is indeed good advice for *all* of our infrastructure, not just flood defence assets.

Another major influencer on how our infrastructure will help us cope, or not, over the next fifty years is the organisational structures we have created for our major resources and associated utilities. Our separation of different functional parts of our support systems is not helpful. Take water as an example; we have public and private bodies dealing with water supply, wastewater treatment, the quality of the water environment, some on catchment boundaries, some on Local Authority

**How to cite this book chapter:**
Venables, J. 2013. Planetary pressures. In: Bell, S and Paskins, J. (eds.) *Imagining the Future City: London 2062.* Pp. 45-47. London: Ubiquity Press. DOI: http://dx.doi.org/10.5334/bag.f

boundaries (both county and local), with different legislative regimes. Decision boundaries are not always drawn in the interests of the planet, the water system or society, but in the interests of the organisations involved.

Will we have, by 2062, an organisation responsible for the water cycle? I do hope so, but I fear not. Perverse investment incentives for water companies do not help. Rainwater is too valuable a resource to end up in sewers. So we need to have created separated sewer systems to capture the water we need for not only potable supplies but for energy production, for growing food and for environmental needs.

Currently, the world wastes one third of the food grown, despite hunger in many places across the UK and across the globe. In fifty years' time, will we waste less? Or will we grow food in laboratories to meet our needs? That would be a great contrast to the rationing of food and minimal wastage in the UK in the 1950s. Will we revert to rationing to reduce waste? Will wasting food become as socially unacceptable as drinking and driving and smoking in public places has now become?

Energy supply, especially of electricity, underpins our current way of life – heating, lighting, air conditioning, transport, and especially communications and control equipment. And much of that energy is still hydrocarbon based, and therefore emits $CO_2$ and other greenhouse gas emissions. If we are to create a low carbon economy and enable low carbon living, then we must use hydrocarbon based energy wisely: to create lower carbon sources, and to make more energy efficient buildings, vehicles and infrastructure. Perhaps most crucially, we need to attune the expectations and behaviours of populations worldwide to the imperative of low carbon living. Two crucial technologies needed to effect the change are dramatic developments in energy efficiency and battery technology, to change our production, delivery, storage and usage of electrical energy.

How will we make our journey to 2062? A look back to the science fiction fifty years ago shows that – through human creativity and innovation – many of those fifty-year-old science fiction ideas are now a reality. Developments from the past fifty years in my lifetime include computers; space exploration; communication methods – small and *very* smart mobile phones, emails etc; food – from rationing to plenty to needless waste; and extensive and cheap travel – the list is extensive.

Although some of the last fifty years of innovation and inventions are in part causes of our current difficulties, many have helped enormously to improve quality of life and the environment. So I think we can and should be hopeful that another fifty years of that human creativity and innovation, this time also spurred on by the climate, carbon and environmental imperatives, can give us the necessary technology developments, whether part of current science fiction thinking or not. Whether better governance and adjusted human expectations and attitudes can be delivered in the same period, we may be less optimistic; but we should remain determined to press for change for the greater good.

Children born in the first decade of the twentieth century will be in their fifties and sixties in 2062, 100 years after Rachel Carson wrote Silent Spring. We wonder what will life be like for them, but also recognise that they will have had at least 25 years of their own careers to shape their own future. Let us hope that we are able to inspire both the current and next generations to reverse the excesses of recent decades, restore the environmental damages those excesses have wrought, and create the necessary sustaining technologies and political governance systems.

# Infrastructure

## Jeremy Watson

Will we recognise London in 2062? Yes, but it will be different in many ways. Since 2018–2020, infrastructure has been designed using holistic engineering principles, and the synergies of co-routing, construction, operation and maintenance are delivering strong benefits of cost and functionality. Waste, water and energy links run in segregated subsurface trunking (operated as a commercial value aggregation service), with built-in condition monitoring to give early warning of leaks or other failures. The system water shortage in the South East was alleviated with the national water grid (still in development), part of which uses the old canal system, which with the addition of pumping stations, proved a surprisingly low-cost solution. London still has one of the best public transport systems in the world, with autonomous guided P2P above ground and an upgraded Underground, with long, fast trains and climate control. Heat extracted from Underground tunnels is sold to business consumers.

### Energy

The Climate Change Act ultimately led to mass adoption of zero-carbon electricity generation and an almost complete move to electrification of the UK's domestic, industrial and transport energy consumption. The country now has a fleet of 30 nuclear power plants, which increased by 10 in the 2030s, after off-shore wind was found to be expensive and unreliable. Local renewable generation is widespread, with over half of homes having either solar-thermal or PV panels, together representing some 30GW. Significant investment in energy storage systems has mitigated the intermittency of renewables. Electricity supply companies, following the 'smart electricity' initiative with dynamic tariff and appliance control functionality, have flexible control of local generation as well as consumer loads. Grid networks were reinforced in the 2020s, with coastal high-voltage DC links routed the length of the country through the North Sea, and with further links to Scandinavia and the Continent. Supergrid pylons are now an unusual sight.

**How to cite this book chapter:**
Watson, J. 2013. Infrastructure. In: Bell, S and Paskins, J. (eds.) *Imagining the Future City: London 2062.* Pp. 49-52. London: Ubiquity Press. DOI: http://dx.doi.org/10.5334/bag.g

## Transport

All private cars are electric (see electric cars), and the development of high-density flow batteries means that fuelling can be done by buying liquid electrolyte at roadside stations or by charging at home. Range is no longer a problem, with a full tank typically offering 1000km or more. Ownership is by leasing, as it is not cost-effective to buy a car, following introduction of the 40% Consumption Tax, which encourages servitisation of capital consumable products. Most families have their own cars, but these are not used in London where Point2Point provides affordable, profiled point to point transport, with subscriber journey aggregation.

Intercity transport is exclusively by the electric HSNet, routinely operating at 350km/hr, with most main cities connected. Cross-city and high-speed links opened up commuting possibilities, making Oxford only 30 minutes from the city.

Air travel is still widely used for international journeys, as biosynthetic avgas is now readily available and affordable. Marine transport is increasing using automated sail technology, especially for freight; this has been driven by carbon emission controls and fuel costs.

## Housing and buildings

In London, housing styles changed radically in some areas and less so in others. Deep retrofit was applied to all old stock by 2050, and now just 50% of the properties extant in the early 2000s are still in use, but with radically reduced energy consumption. Retrofit proved very difficult, and the UK did not meet its targets until 2055. In many cases, multiple phases of renovation were needed. The non-heritage scrappage scheme enabled replacement of buildings beyond economic renovation, and has allowed the City to be remodelled into a high-density, mixed-use environment. Building Information Modelling (BIM) is now ubiquitous for capital construction in the public and private sectors, applying from design visualisation and co-creation with clients through computer-aided off-site manufacture, to real-time performance optimisation.

## Food and retail

Supermarkets still exist, but they sell only food. The processed elements of much of this are manufactured on-site in hyper-automated factory modules. Consumer retail is almost exclusively via the web, although high-value clothing is customised, with measurements taken and delivered on-line via high street stores. Where shops exist, they are essentially showrooms for web goods. The high street is now mainly a meeting place with a large variety of eating places and personal service providers.

What used to be capital consumer purchases are now provided as services, with the provider owning the asset and maintaining it to the latest and best standards. 40% VAT on capital consumer purchases has essentially enforced leasing as a model for such commodities.

## Health, wellbeing and aging

By 2020, the National Health Service was faltering and private wellness insurance and the stabilised aging demography were driving personal behaviour. Home monitoring, assisted living and affordable personalised medicine are now everyday facts of life. Telehealth and telecare are offered by real-time service providers on a subscription basis; some of these providers are subsidiaries of health insurers. Other real-time services are often 'bundled' with this – security and home energy control being typical. Personalised medicine means that most tissues can be replaced with cultured stem-cell derived material, harvested from the individual earlier in their lives. Options are provided in wellness insurance for ceasing interventions at a certain age, this has led to extended quality life followed by very rapid decline, and this is frequently planned and managed. It is usual for people to die (see death) in their own homes, so public care homes are the exception rather than the rule. Home automation is as commonplace as entertainment systems; frequently this includes mobile house avatars, providing assistance and companionship.

## Security

Identity theft and financial fraud had been growing problems until the introduction of DNA validation. This is now used in most major transactions. The elimination of cash also reduced street crime. Cybersecurity continues to be a major concern, but netbot policing has reduced this to manageable levels. However, increased teleworking has given rise to the 'cyber-employee' problem, where staff members defraud employers by using computers to simulate work attendance, giving new relevance to the 'Turing Test'.

## Education

Teacherless schools were introduced in the 2050s. Schooling now takes place through distance learning and tele-tutorials, with physical attendance supervised by para-tutorial staff who focus on sport and social skills.

Universities have long since clustered into virtual Centres of Excellence, and close partnerships with business have meant that learning experiences are closely tailored to the needs of individuals and future employers. A permeable approach to further and higher education has encouraged apprenticeships and made transfer to degree courses readily available for students of all ages. More than half of undergraduate courses are now delivered at large scale by distance learning, following the adoption of Massively Open Online Courses in 2012–2020. Virtual tutors allow one-to-one and group support using AI technologies.

## Information

Ubiquitous ICT led to most people suffering information overload by 2020. Smart Personal Algorithmic Logic now contextualises and bundles information so individuals only see what is relevant to their location and current activities. High-speed wireless communication in urban areas has enabled all but the most elementary everyday device to have its own web identity, to report its location, state of health and current usage status. The world has become 'object-orientated' with physical items holding and communicating design, maintenance and status information – an essential requirement of the servitised economy.

## Economy

Growth is small but positive following negative population growth and the radical policy overhauls which were dictated by the failure of classical economics in the '20s. Most employment is in the service sectors, with hospitality, entertainment and maintenance strong. Manufacturing is now re-localised and hyper-automated, yielding strong industrial profitability; this has strengthened tax revenues enabling adequate welfare benefits, although most public sector activity is executed through GovNet, and the civil service is 10% of its size at the beginning of the 21st century.

# Things

Water, air and food are the stuff of life. Without housing, energy and transport, living in the city would be impossible. *Things* addresses the physicality of London, the systems and materials that make up the built environment.

By 2062 human-induced climate change will be a key consideration for London; its impacts on the environment of the city will be far more evident than they are today. All of the chapters in this section anticipate these changes, addressing climate change as it relates to a range of urban systems. The chapters include scientific assessment of likely future impacts of climate change, as well as the key policy and design measures for mitigation and adaptation. If serious efforts are made to reduce carbon emissions, we will see changes to the ways in which energy is produced and consumed. The contributions address the scale of measures needed to address climate impacts and to achieve emission reduction targets. It is clear that if London's leaders wish to secure a sustainable future for the city, they must make some bold decisions in the coming decades.

The Mayor of London has presented an ambitious target for carbon emissions: by 2025 emissions from all sectors should be only 60% of 1990 levels. This reduction can only be achieved through strong investment, changes in land use and planning, and extending policies that discourage car use and improve energy efficiency. This target presents a fundamental challenge to the current governance of London's transport and energy. The scale of the measures required is incompatible with a political discourse that excludes consideration of options that go beyond incremental change; serious consideration should be given to options that challenge car-dependent transport systems, or centralised energy networks.

Climate change is not the only driver for radical transformation. For instance, reducing the dependence of London's transport and energy systems on fossil fuels would improve both air quality and energy security. If London has made the transition to a low carbon, sustainable economy by 2062, it will be the result of changes in governance and institutions, as much as technologies and individual behaviour.

Future proofing London's buildings and infrastructure begins with a good understanding of the likely impacts of climate change, and leads to new design techniques and technologies. Many of the buildings in London in 2062 have already been built, and techniques for retrofitting to reduce emissions and adapt to a changing climate will be crucial. London buildings must be better insulated to reduce energy demand, and better ventilated to reduce over-heating. Buildings will be designed to be more resilient to flood, to reduce runoff to sewers and to minimise their demand for water. By 2062 it is likely that Londoners will be drinking recycled water, as well as collecting and recycling more of their own water for toilet flushing and garden watering. The Thames Barrier should still be providing adequate protection against tidal surge flooding, but planning and construction for its replacement should be well underway.

By 2062, London could be producing a much higher proportion of its own food, and have reconfigured its relationships to the farmers who supply its food from farms elsewhere in the UK, Europe and the rest of the world. Like all the changes in basic urban functions, producing more food in London is likely to be accompanied by social and political change, visible as a more diverse and edible urban landscape.

# Transport, climate change and society

Robin Hickman

## Transport and climate change

London faces huge challenges in moving to a reduced reliance on oil and lower carbon dioxide ($CO_2$) emissions in its transport system and travel behaviours. The city is one of the most progressive internationally, at least in policy terms, in seeking to respond to climate change. It has a very high public transport mode share for radial trips to the central area, but is still heavily reliant on the private motor car (almost exclusively powered by petrol or diesel) for travel in the suburbs[1].

The Mayor of London has set a very demanding target on climate change: a 60% reduction in $CO_2$ emissions by 2025, across all sectors, on a 1990 baseline (Greater London Authority, 2007). Although there is no specific target in the transport sector, the expectation is that transport should make a significant contribution. The scale of the target set is consistent with the aspiration to limit the increase in global surface temperatures to 2°C, at most, relative to present day levels (Intergovernmental Panel on Climate Change, 2007; Stern, 2009). There is, however, less certainty as to how to achieve this target in terms of the level of investment required and the ability to influence individual travel behaviours. The response to climate change also neatly correlates with the oil scarcity problem. Many of the solutions required, such as a much greater use of public transport, walking and cycling, also mean there is likely to be less reliance on the petrol-fuelled motor car as the main mode of travel. The International Energy Agency (2009) points towards the urgency of the problem: there are 'just

---

[1] For Greater London as a whole, the car accounts for 40% of daily journey stages, with most car usage in outer London. Walking accounts for 21%, bus 18%, Underground 10%, and cycling 2% of journey stages. Public transport is, of course, very well utilised for commute trips into central London. Journey stages are defined as follows: a journey trip is a complete one-way movement from origin to destination by one person for a single purpose, comprising a number of 'stages', e.g. walk to the station, Underground trip, and walk to work is one trip and three stages (Transport for London, 2007).

---

**How to cite this book chapter:**
Hickman, R. 2013. Transport, climate change and society. In: Bell, S and Paskins, J. (eds.) *Imagining the Future City: London 2062.* Pp. 55-62. London: Ubiquity Press. DOI: http://dx.doi.org/10.5334/bag.h

46 years left' of conventional oil consumption, assuming proven reserves and current consumption rates. Though there are additional and large reserves of unconventional oil, surely we should be considering in earnest what our transport futures could be like under very different external conditions. This chapter considers these issues in terms of the potential transport scenarios possible for London to 2062, and the likely wider governmental and societal changes required if travel behaviours are to become less carbon intensive to any significant degree.

## Transport scenarios to 2062

Scenario analysis allows the comparison of potential policy trajectories, including possible trend-breaks away from the 'business as usual' (BAU) pathway to something more akin to a sustainable travel future. Each scenario includes a variety of policy measures, including different levels of 'application' (covering policy initiative, investment and approach to implementation). Scenarios can be developed by considering the major drivers of change affecting transport in London. These are given in Table 1. Some are fairly certain in outcome, or at least well researched and understood as to their likely impact (e.g. demographic change). Others are more uncertain (e.g. the extent of increased environmental awareness and changed behaviours in the population). Many of the drivers have conflicting impacts in terms of likely travel distance, mode share and consumer choice.

Many of these trends can have dramatic impacts on travel patterns in London in future years, certainly over the long term to 2062. Figure 1 illustrates transport scenarios in London using a classic scenario 'dilemma' matrix (drawing on the approach from Schwartz (1996) and Van der Heijden (1996)). Two of the major issues from the drivers of change, in this case the extent of technological change and environmental stewardship, are used to generate the dimensions of change within the scenarios. These are chosen to highlight the major problem currently facing transport planners in London (and indeed the wider UK and international arena): whether the gains in vehicle efficiency with cleaner vehicles can allow us to remain as mobile as we are, or become even more mobile, with use of a similar mode share; or whether travel behaviours need to change markedly as well as the vehicles. Do we need to rely on more than technological change, and to think through how to change travel behaviours more effectively than we do at present?

| Emerging Socio-Demographic Trends | Potential Travel Implications |
|---|---|
| • Changing demographic and household structures<br>• Increasing world trade and globalisation<br>• Economic volatility, including periods of financial collapse and recovery<br>• Rising importance of local activity provision, but greater discernment in choice of activities and consumer purchases<br>• Rapid technological developments and the emergence of 'digital natives' (the new generation growing up accustomed to the use of technology)<br>• Taxation increasingly based on resource and energy consumption rather than income<br>• Decline in the power of national governments and distrust in institutions (with a reduced ability to influence change)<br>• Increasing awareness of sustainability issues and demand for change in opportunities and lifestyles | • Steady increase in demand for mobility, particularly long distance and with speedy modes – for passengers and goods; with periods of reduced growth related to economic volatility<br>• Huge growth in demand for public transport, walking and cycling, as environmental and health benefits become much more widely known and sought after<br>• Increased importance of and attention to the quality of the journey experience<br>• Aggregate travel times remain steady<br>• Gradual increase in share of low emission vehicles and use of alternative fuels<br>• Increased trip distances in goods movement; though partial reduction through localised sourcing<br>• Much greater realisation of the 'Network Society' – electronic flows replace part of physical travel as well as opening up a new range of social interactions |

(Developed from the Department for Trade and Industry and Office of Science and Technology, 2006)

**Table 1:** Drivers of Change.

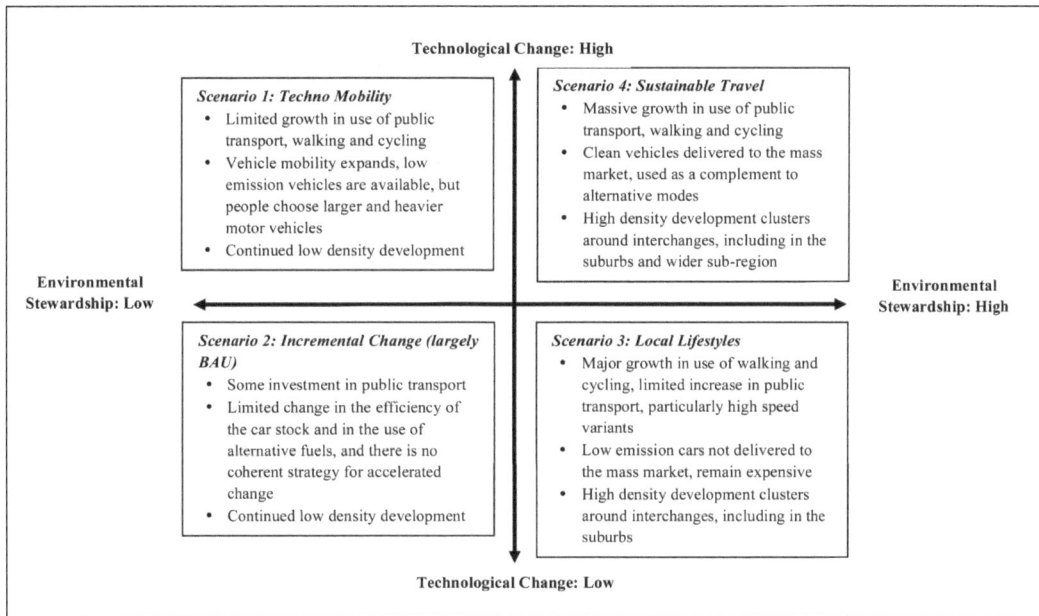

**Figure 1:** Potential Transport Futures 2062.

## Governance and societal implications

Clearly there are multiple potential futures in transport, four of which have been described above. Some are more preferable than others. All, with perhaps the exception of the current incremental change 'BAU', are likely to prove difficult in implementation terms. This is a major problem politically: governments have notorious difficulties in moving public behaviours away from the mainstream trends. The recent political progression in the UK, at the national level, has been towards a neoliberal viewpoint. Individual 'choice' is protected as the primary interest, and 'governmental intervention' perceived as difficult and unwarranted. For example, there is a tendency to rely on the 'nudge' in policy making, drawing on the work of Thaler and Sunstein (2008), rather than more interventionist approaches. This strategy, of course, is very unlikely to reduce transport $CO_2$ emissions to any significant degree. This key point is either overlooked or ignored, depending on the acceptance of the RealPolitik and Greenwash[2] reading of events.

In London, an ambitious policy and investment programme has been developed in transport, with a series of major transport projects developed. But, even where the ambition is greater, where there is a seemingly well developed future strategy, there are often large difficulties in implementation, both in terms of political deliverability and public acceptability. These issues can be explored using the concepts of policy discourse and 'discourse coalitions'. The current dominant governmental process in London can be viewed as reflecting the 'environmental management' approach, where the sustainability problem is seen as a fundamental failure in the workings of the institutions of society, yet the solution is seen as a combined 'techno-institutional fix', and importantly using the current institutional arrangements, i.e. a little better regulation will suffice (Hajer, 1995).

---

[2] 'RealPolitik': the politics and policy development based primarily on power and on practical considerations, rather than valued notions or ethical premise. The term can be linked to the practice of 'Greenwashing', whereby the conjecture in policy making concerns environmental sustainability, but little is ever done in terms of actual investment and policy initiative.

This is a narrow conceptualisation of the potential options for policy delivery, of power and power relations, and different potential pathways to achieve societal goals. The framing and definition of the problem and solution is thus very important.

The more critical observers argue that the mainstreamed use of 'sustainability' as a concept has been a 'rhetorical effort' concealing a strategy for sustainable development, but with the emphasis strongly in supporting gross domestic product (GDP) growth and profit maximisation, at the expense of environmental and social policy goals, and ignoring the basic contradictions that are evident within the sustainability concept. Thus, the conventional storyline attempts to 'reconcile the irreconcilable' (in this case, the environment and development) (Hajer, 1995; Jackson, 2009). These types of issues are evident in the transport sector in London. The London Plan offers a pro-growth paradigm, and there is little effort to radically reduce travel distance and change mode share in London, particularly in the suburbs. Individuals and organisations within London and the UK have widely differing views as to what 'sustainable travel' or 'sustainable cities' might mean. Take, for example, the King Review (King, 2008) and the Eddington Study (Eddington, 2006), where the focus was on improving the delivery of low emission vehicles, supporting long distance travel (by air, motor car and high speed rail); or the environmental group lobby, where the call is for a much greater use of short distance travel, inter and intra-urban based public transport, and walking and cycling. Many of these actors postulate that they are trying to achieve greater 'sustainability' in travel behaviours. But there is often little in terms of shared values or common interests, and people are working to different definitions of the 'sustainable travel' concept.

So we have a messy and confused understanding and application of the sustainability concept in transport. Few actors in the sustainable transport field appear to have developed a radical social critique; it is assumed that very radical targets can be achieved within the same institutional system. However, this is far from clear from the evidence and the current trends. The transition to sustainable transport does not appear to be a 'value free' process of convincing various actors of the importance of 'the green case'. The history is much more of a 'struggle' between different coalitions of interests and beliefs, each consisting of and supported by scientists, politicians, activists, research groups, lobbyists, marketing and advertising groups, newspapers, film, TV, journals, trade press, and even celebrities. If examined closely, the various groups are very fragmented and contradictory. Each group develops and supports its own particular discourse, a particular way of thinking and talking about sustainability. Some develop momentum and influence, others fade from view (Hajer, 1995). At the individual level, we also act upon our own 'images of reality', drawing on certain discourses that we engage in or are engaged with. We may think that our choice of travel and its impact on the environment is limited, that our lifestyles are fairly fixed, that it is 'too difficult' to move away from car based mobility given complex lives, that public transport is too expensive to provide to the mass market beyond the urban centres, and that we can do little to reduce $CO_2$ emissions in the transport field. This is certainly the dominant discourse, but there are of course many others, some differing in view and action, and some significantly so.

In London, the policy measures being employed are relatively progressive, and compare very well to practice elsewhere internationally. There is a congestion charge (introduced in 2003), albeit for a small part of central London (and reduced in area coverage since the removal of the western extension), charging £10 for vehicles to enter the cordon. There are many impressive investments in public transport, either recent or planned, such as the Jubilee Line extension (opened 1999); London Overground, an orbital rail link around suburban London (the first sections were established in 2007); some state-of-the-art interchange redevelopments, such as King's Cross/St Pancras (the refurbishment opened 2012); and Crossrail, a major new east-west railway link across central London (scheduled to open in 2018). However, even with these measures, it is unlikely that transport $CO_2$ emissions in London will reduce by anything near 60% by 2025 on 1990 levels. The current Transport for London approach is only likely to reduce transport emissions in the order of 20–25% (Hickman et al, 2009). The critical 'discontinuity' measures, i.e. those that make a significant difference, are not being implemented to any degree. Perhaps this is with good reason: there is little public understanding of the scale of change required to reduce transport $CO_2$ emissions by 60%, or deeper to 80% and, perhaps as a result, little political appetite.

To explore these issues further, the political possibility of delivering particular policy measures can be examined in terms of the degree of public authority (legitimate coercion) required to implement them (Dunn & Perl, 2010). The level of 'coercion' is viewed as reflecting an element of 'forcing' another party to behave in a non-voluntary manner (through action or inaction). A 'weak state tradition', as increasingly experienced in the UK, means that policy tools are generally chosen from the least coercive part of the spectrum. Regulation on businesses and individuals results in hostile responses, intense lobbying and the 'watering down' of original proposals. Increased taxes, for example, begin to be viewed almost as 'legalised extortion'. Subsidies and voluntary schemes prove much more acceptable to politicians and the public, but of course only if maintained at a limited scale (Dunn & Perl, 2010). Again, this framing of the debate is critical. Many of the measures that would make a difference to reducing transport $CO_2$ emissions are also the ones that are perceived as undeliverable politically in the current context. There is little chance of road pricing across the Greater London area. There are no great moves to increase densities in the suburbs. There are no plans to develop tram or bus rapid transit schemes in Outer London, to provide segregated facilities for cycling or a new series of public realm improvements, or to encourage with subsidy a large number of low emissions vehicles in the London fleet. All of these types of radical measures would be required to reduce $CO_2$ emissions to a significant degree in London, to levels reaching the ambitious 60% reduction target.

The risk society thesis (Beck, 1992) suggests that the basis on which environmental politics have so far been made has to be fundamentally rethought. Climate change is not simply a problem that can be regulated away, and certainly not by the existing institutional arrangements. These have been successful in producing unprecedented wealth for a small cohort of the population, but this has occurred only with considerable social and environmental cost. Beck (1992) argues that the ecological crisis might become the 'stepping stone' to a new and superior form of modernity. But the existing institutions are 'digging their own grave' by increasingly showing their inability to handle the dangers they have themselves produced. Though London is progressive relative to most other contexts, the Beck thesis still seems to stand. Many of the policy approaches being considered, and implemented, are having little impact against the strategic targets being set at the city level. Almost all public observers demand more progress in moving towards lower $CO_2$ emissions. Scenario 4 (Sustainable Travel), as developed earlier, is perhaps what policy makers would like to deliver, but it is not at all clear that this level of environmental stewardship, even in London, is being achieved. The level of investment in public transport and the other non-car modes has not changed to the expected degree. In the end, we do not seem to be moving far away from Scenario 2 (Incremental Change).

This intractable conflict is hidden in the initial definition of the problem, in the issues discussed and those that remain undiscussed. The framing of the debate makes certain elements seem 'fixed' and inappropriate, others are viewed as 'problematic', and some much easier to discuss and 'deliver' (Hajer, 1995; Hickman & Banister, 2013). The classic examples in transport are measures such as road pricing, reducing space for private cars, increasing densities in suburban areas, and, of course, reducing the growth in international air travel: all of these are seen as 'politically difficult' and usually remain beyond the mainstream debate. These are debated as seemingly 'technical' positions, but of course conceal a normative stance, supported by the institutional arrangements (Hajer, 1995). The contest between development, travel and sustainability is often an ideological one, rooted in fundamentally different value systems and worldviews, and is closely linked to our consumption-based society (Urry, 2007). It is unlikely that we can meet $CO_2$ reduction targets and aspirations to reduce climate change whilst pursuing the consumerism model (Wheeler, 2012). This is inherently problematic and does not seem to be understood or even debated. Debord (1967) asks that we 'wake up the spectator who has been drugged by spectacular images'. This appears very relevant still: there is an opportunity, but it is a narrowing window of opportunity, and perhaps, as yet, there is little appetite to tackle the fundamental issues to any significant degree. Much like Plato's *Allegory of the Cave* (Plato, ~380), we have lived chained to our car and oil dependency for the major part of our lives. We have ascribed forms to the lifestyle, believing that it brings freedom, status and even expression, but in reality there are many hugely adverse impacts that we continue to ignore. We (simply) need to realise that there are much more attractive travel lifestyles on offer: who wouldn't want to commute, Copenhagen-style, by bicycle; to use the French-style trams in the suburban centres; to use our electric hybrids whenever we need to use the car? We just need to look outside to the sunshine – but it seems the real debate is very slow to develop: to consider how transport is important to the development of the city and to society itself, and how the shaping of society and its institutions affect our travel.

# References

Beck U. 1992. *Risk Society: Towards a New Modernity.* London: Sage

Debord G. 1967. *Society of the Spectacle* [first published as *La Société du Spectacle* by Buchet-Chastel, 1967]. Eastbourne: Soul Bay Press. Reprinted 2009

Department for Trade and Industry and Office of Science and Technology. 2006. *Intelligent Infrastructure Systems. Project Overview.* London: DTI and OST

Dunn J. and Perl A. 2010. *Launching a post carbon regime for American surface transportation: Assessing the policy tools.* Lisbon: World Conference on Transport Research

Eddington R. 2006. *The Eddington Transport Study* [Rod Eddington]. HM Treasury. Norwich: The Stationery Office

Greater London Authority. 2007. *Climate Change Action Plan.* London: GLA.

Hajer M. 1995. *The Politics of Environmental Discourse: Ecological Modernization and the Policy Process.* Oxford: Clarendon Press

Hickman R, Ashiru O. and Banister D. 2009. *Visioning and Backcasting for Transport in London (VIBAT-London). Stages 1-4, Background Reports.* London: Halcrow Group

Hickman R. and Banister D. 2013. *Transport, Climate Change and the City.* Abingdon: Routledge

Hood C. 1986. *Administrative Analysis: An Introduction to Rules, Enforcement and Organisations.* New York: St Martins Press

International Energy Agency. 2009. *Transport Energy and CO2.* Paris: IEA

Intergovermental Panel on Climate Change. 2007. *Fourth Assessment Report on Climate Change. In: Synthesis Report.* Geneva: IPCC

Jackson T. 2009. *Prosperity Without Growth: Economics for a Finite Planet.* London: Earthscan

King J. 2008. *The King Review of Low Carbon Cars [Julia King]. Part 2: Recommendations for Action, HM Treasury.* Norwich: The Stationery Office

Plato. ~380 BC. *The Republic.* Reprinted 2007. In: Lee H. and Lane M, editors. London: Penguin

Schwartz P. 1996. *The Art of the Long View: Paths to Strategic Insight for Yourself and Your Company.* London: Currency Doubleday

Stern N. 2009. *A Blueprint for a Safer Planet: How to Manage Climate Change and Create a New Era of Progress and Prosperity.* London: Bodley Head

Thaler R. and Sunstein C. 2008. *Nudge: Improving Decisions About Health, Wealth, and Happiness.* New Haven: Yale

Transport for London. 2007. *Travel in London*. London: Transport for London
Urry J. 2007. *Mobilities*. Cambridge: Polity
Van Der Heijden K. 1996. *Scenarios: The Art of Strategic Conversation*. Chichester: John Wiley
Wheeler S. 2012. *Climate Change and Social Ecology: A New Perspective on the Climate Challenge*. London: Routledge

# Decentralising energy

## Peter North

The Mayor of London's Climate Change Mitigation and Energy Strategy sets ambitious targets to reduce London's $CO_2$ emissions to 60% of 1990 levels by the year 2025. With 30% of the capital's emissions attributable to heating, mostly from mains gas, one of the greatest opportunities is to reduce demand for heat through building retrofit and low carbon, local (decentralised) heat supply by means of Decentralised Energy (DE[1]). Decarbonising the other big energy related emitter, electricity supply, is best placed as a national action through nuclear, wind and carbon capture and storage.

The Mayor recognises the importance of decentralised energy technology in contributing towards the carbon dioxide ($CO_2$) emission reductions and sets a further target of supplying 25% of London's energy supply from DE by 2025. Current Greater London Authority (GLA) planning policies require relevant developments to consider: a) connecting to local district heating networks and if not, then b) installing their own Combined Heat and Power (CHP), and c) meeting 20% of the site energy demand from renewable energy sources. The current uptake of DE falls short of the trajectory required to meet the 2025 target, yet work carried out by the Greater London Authority concluded that London does have the capacity to deliver the targets. So the market is failing to deliver the potential which Powering Ahead (Greater London Authority, 2009) estimated to be worth £5–7 billon of investment to deliver annual $CO_2$ savings in the range of 2.2 to 3.5 million tonnes.

During 2007, the London Development Agency (LDA), the Mayor's now abolished delivery agency, investigated the DE market failure as part of the London Thames Gateway Heat Network (LTGHN) project. It concluded that the largest quantum of $CO_2$ savings could be delivered at market competitive rates (i.e., without government energy subsidy) in dense urban areas through industrial scale combined heat and power involving extensive district heating (DH) networks. This role for DH networks is mirrored in the Government's Heat Strategy which also came to a similar conclusion following its own work. Published in March 2012, the Heat Strategy highlights

---

[1] Decentralised Energy – defined as the local generation of electricity and where appropriate, the recovery of the surplus heat (combined heat and power – CHP) for purposes such as building space heating and domestic hot water production.

---

**How to cite this book chapter:**
North, P. 2013. Decentralising energy. In: Bell, S and Paskins, J. (eds.) *Imagining the Future City: London 2062.* Pp. 63-66. London: Ubiquity Press. DOI: http://dx.doi.org/10.5334/bag.i

a potentially major role for networks in areas of high heat demand by removing carbon from commercial and domestic building heat supply.

A city-wide DH network such as the LTGHN could be viewed as creating a heat market, by providing a route to market for low and zero carbon heat suppliers (industrial undertakings such as energy from waste, combined cycle gas turbine plant and energy intensive industry), and to heat consumers for building space heating and domestic hot water (DHW) production and other heat requirements. The LTGHN project was estimated to cost £160 million (at 2008 prices), serve the equivalent of over 110,000 homes and principally involved the creation of a district heating network that would involve:

- The phased construction of seventy km of DH pipework over a ten – fifteen year period connecting private sector industrial plants to consumers (a small to medium sized system in European city terms)
- The buying of heat from industrial undertakings
- The selling of a secure[2] supply of heat to consumers in the form of hot water
- The operation and maintenance of the heat network
- The expansion and development of the heat network and new connections
- Where appropriate, the facilitation of the bilateral supply of heat between suppliers and consumers, by allowing system access and charging 'use of system' for the transfer of their heat.

The project development was suspended in 2010 following the failure of the private sector to respond to a formal invitation to negotiate heat supplies. Considerable strategic and technical know how and commercial principles were established that continue to be deployed on London's current DE developments. The LTGHN would in fact look very similar to the city-wide CHP district heating schemes that have been operating in Northern European cities for many decades. Most of these schemes were undertaken by the municipality for reasons of national energy security (Denmark), the most economic form of urban energy supply (Finland), or the efficient utilisation of energy. In contrast, the LTGHN project was predicated on $CO_2$ savings at market competitive heat prices. Despite being economic propositions, the delivery of projects on the scale of LTGHN could be greatly enhanced by the de-risking effect of initiatives from government to facilitate the involvement of local authorities in promoting heat networks at scale, to encourage connections of heat sources and heat loads to these networks and coordination between local authorities. The implementation of the Government's Heat Strategy therefore provides an opportunity to secure this.

The development of large-scale DE in London continues primarily through the activities of London Boroughs following a systematic methodology of local policy formulation and heat mapping to identify the most heat-dense areas, and energy master-planning to establish the evidence basis and high level costs of specific area-wide DH systems. Local authorities (LAs) are then able to deploy their powers to de-risk projects by requiring developments to investigate connecting to the DH network, require financial contributions from new developments and facilitating other 'buy-out' arrangements, bringing forward their own heat loads to secure long-term heat income for the project and possibly funding at public sector rates. Similarly, LAs may also have an interest in existing and proposed energy from waste (EfW) schemes and can require new energy developments to be built with heat off-take to supply local DH network. With a number of London's DE projects currently completing the feasibility stage and moving towards commercialisation, it will be interesting to reflect on the public sector role in their delivery.

The question has been asked, 'So what happens when the gas runs out and we're so efficient at recycling, there is no longer any waste?' By then, interconnected DH networks would have been established in the short-term from gas CHP, in the medium term larger schemes based on EfW and

---

[2] Heat is made 'secure' with the addition of standby and top-up heat-only boilers fuelled by mains gas in sufficient number and capacity to meet heat demands in the event of the industrial heat supply failing.

sources of surplus heat, and finally the networks interconnected to form city-wide systems with multiple heat suppliers. As DH networks have the simple role of circulating hot water as the energy carrier, they are heat-technology agnostic. They don't care where the heat comes from, so it is entirely feasible that when the gas runs out and there is no longer any waste, the future energy source will be electricity from nuclear power and wind, CHP would therefore be replaced by industrial heat pumps.

But there are other more efficient and effective possibilities. Further consideration of alternative city-level energy sources has found there to be considerable potential in low grade (temperature) surplus heat from the likes of data centre cooling, underground train ventilation, electricity substations, sewage works etc. Heat pumps can elevate this low grade heat to current day DH supply temperatures (70°C – 110°C), however the higher the temperature elevation, the lower the heat pump efficiency and there is a limit to the economics and what people are willing to pay for heat from such a system.

By way of example, the elevation of waste heat from an underground train vent (23°C – 28°C) would be limited to 55°C – 60°C using a heat pump so as not to exceed the market price of heat. This would relate to a coefficient of performance (CoP)[3] of around three. In fact, lower supply temperatures may be entirely viable if building heating systems could either accommodate the lower supply and return temperatures, or are designed for this from the outset. Taking this to a natural conclusion, why bother with the heat pump? Why not simply collect and supply heat to the DH network at low temperature and elevate the temperature to the end user requirements by a local heat pump at the point of consumption? Even the need for a heat pump for the user could be minimised or eliminated if the building heating system was designed for low temperature, ie underfloor heating or close-coupled wall heating. Or maybe the traditional wet radiator system operating at low temperature would be sufficient where properties are highly insulated following retrofit.

So the future London urban DH network will evolve today from natural gas CHP, energy from waste and surplus heat operating at higher temperatures, with the networks becoming more interconnected. The systems would mature into low temperature networks scavenging low grade surplus heat, minimising the need for primary energy input. The system will be very efficient due to the low system heat loss, and the local distribution legs to consumers will be cheaper because of less onerous pipe work material requirements, with any high temperature water requirement being met by a local heat pump. Regulation should require industrial and commercial cooling systems to be designed to connect to low temperature DH systems, and the use of air radiator/river water cooling discouraged. Inter-seasonal aquifer heat storage would also become a possibility. Such systems have already been thought about and exist as small campus-type systems where it has been possible to carry out the overall design and specification from production to consumption in a single system. So is this how a zero carbon London will be heated in 2062?

## References

Greater London Authority. 2009 (October). Powering ahead – Delivering low carbon energy for London. Available from: http://legacy.london.gov.uk/mayor/publications/2009/docs/powering-ahead141009.pdf. [Accessed 6 August 2013]

---

[3] The coefficient of performance (CoP) of a heat pump is the ration of heating or cooling provided over the electrical energy consumed. The CoP provides a measure of performance for heat pumps that is analogous to thermal efficiency for power cycles. The equation is:

$$COP = \frac{Q}{W}$$

where $Q$ is the heat supplied to or removed from the reservoir and $W$ is the work consumed by the heat pump.

# Taking carbon out of heat

## Bob Fiddik

If government targets are to be met, by 2062 all London's homes and workplaces will have been virtually zero carbon for twelve years. Currently the energy used by London's buildings is responsible for 80% of the city's carbon dioxide ($CO_2$) emissions. Almost 50% of these emissions arise from demand for heating and hot water. The majority of this demand is currently met through the national gas network supplying individual gas boilers in homes and workplaces. Taking carbon out of heat is also an immense retrofitting problem as around 80% of the buildings with us now will still be in use in 2050.

## Learning how to do policy again

Heat is a relative newcomer to current UK energy policy making, but energy policy making itself had to be relearned after two decades when it was left to the market and a regulator whose only concern was price. It was the growing need to tackle energy security and climate change that led to a return of energy policy at the start of the 21$^{st}$ century. Agreeing the set of high-level objectives was a fairly quick and simple task, these being to secure supplies of energy that are:

- affordable
- secure, from diverse sources
- sustainable and low carbon.

But agreeing how to get there has been neither quick nor simple. UK energy supply is dominated by the centralised, top-down networks for electricity and natural gas supply. In the absence of supplies of 'low carbon' gas, it is easy to see why policy making has focused on mechanisms to get more large-scale, low carbon generation feeding into the grid.

At the other end of the system, policy has tackled demand for heat in new buildings through the Building regulations. For existing buildings, heat policy has been limited to obligations on suppliers to fund insulation and heating system upgrades in dwellings. This funding has been greatly

**How to cite this book chapter:**
Fiddik, B. 2013. Taking carbon out of heat. In: Bell, S and Paskins, J. (eds.) *Imagining the Future City: London 2062*. Pp. 67-72. London: Ubiquity Press. DOI: http://dx.doi.org/10.5334/bag.j

reduced with the introduction of the 'Energy Company Obligation' (ECO) alongside the 'Green Deal' (which is a loan not a grant).

## The 'all electric' detour

Given the difficulties in influencing how millions of consumers use energy, it is easy to understand why supply-based technical fixes for decarbonising heat would be so seductive. For a number of years the prevailing view among policy makers was that the optimum solution was to:

- move heating from gas to electricity – primarily by installing heat pumps
- decarbonise the electricity grid – using nuclear, carbon capture and storage (CCS) and large-scale wind.

But it gradually became clearer that a major obstacle along this route was the highly seasonal pattern of demand for heat. Around 60% of the extra electricity plant that would be required to meet winter peak heat demand would be idle for six months of the year. In addition, transmission and distribution networks would need upgrading to carry the increased loads, along with all the substations. All of these additional costs would need to be met by electricity bill payers.

It was this seasonality that was highlighted as a key issue in policy documents (in particular 'The Future of Heating' (Department of Energy & Climate Change, 2013)). But there are two further major problems facing the electric heat solution which have been more difficult to admit in policy documents:

- the performance of retrofitted heat pumps
- the rate of grid decarbonisation.

## Heat pump performance in retrofits

Heat pump performance (coefficient of performance, COP) decreases with increasing temperature difference between the 'source' (where the pump gets the heat) and the 'sink' (where the pump delivers the heat). So ground source heat pumps can achieve higher efficiencies than air source heat pumps (ASHP) as ground temperature in winter is higher than ambient air. A new build property with under-floor heating supplied at around 45°C will achieve a higher heat pump efficiency than an existing building with standard radiators supplied at 75°C. For urban and suburban retrofits it will be ASHP that would be adopted (due to their smaller space requirements) supplying standard radiators. These installations would need to be accompanied by solid wall insulation (and in some cases oversized radiators) to achieve acceptable COP levels.

With solid wall insulation being too costly to be delivered by the Green Deal mechanism, a large proportion of the ECO was designed to subsidise it. Experience from most community-scale energy efficiency programmes is that take up rates are low even when insulation has been offered for free. The Green Deal has had a very slow start, and the jury is still out on whether it is attractive enough to consumers to deliver the targeted reductions in energy demand. In addition to the economic case, uptake of external cladding may well be limited by conservation areas and homeowners' aesthetics, while internal cladding is thought to involve too much hassle and loss of internal space.

## Grid decarbonisation

Perhaps the most serious obstacle is that the policy milestones for grid decarbonisation are effectively being pushed further into the future by 'realpolitik'. All forms of low carbon generation involve some additional cost, and in tough economic times no government wants to be seen as loading additional costs on to consumers' energy bills. The new 'Contracts for Difference' (CfD) mechanism is designed to drive low carbon electricity generation by guaranteeing a minimum price (so called 'strike price') to generators should the market price be lower. Despite lengthy negotiations with potential nuclear developers, at the time of writing a strike price has not been agreed.

Any cost and delivery time estimates for new nuclear need to take into account the evidence from the only two new build projects in Europe. Both are vastly over budget and have been delayed by many years. With CCS yet to prove its large-scale technical and economic viability, UK electricity looks set to continue to rely on gas for many years (whether from domestic 'fracked' shale resources or imported).

## Then back to the '70s... sort of

We have to decarbonise electricity, but with electricity demand itself steadily growing, it would be crazy to add to this if there were other ways to remove carbon from heat.

In all of the media and public debates about energy there has been little mention of the fact that our centralised power stations reject around two thirds of their input energy as waste heat – in total, roughly the same amount that is needed to heat all buildings in the UK. This is down to the laws of thermodynamics rather than poor design. The majority of our electricity is produced by burning a fuel to heat water into steam which then drives a turbine generator. The greater the temperature drop between the steam entering and exiting from the turbine, the greater the electricity output. So, UK plants optimise the electricity output by using cooling towers or sea water (in the case of nuclear plants) to cool the exit steam (down to around 35ºC).

Following the oil price shocks of the 1970s, Denmark converted their power stations to run as Combined Heat and Power (CHP) plants and to extract the waste heat at a higher temperature (around 110ºC) so that it can be used to supply city-scale district heating schemes distributing hot water to buildings. This results in a loss in electricity output from the power station, but the critical point is that you typically get seven kWh of heat for every kWh of electricity lost (a 'virtual' COP of 7). This beats all practical retrofit heat pump installations (which typically achieve a COP between 2 and 3). And this performance can be achieved without first having to insulate all those solid walls.

Over the following decades, the Danes realised further benefits from having installed these heat networks:

**Fuel flexibility** – Hot water is distributed rather than a specific fuel (like gas), so it has been simple to switch to cheaper or lower carbon fuels. In Copenhagen, district heating covers 98% of the city, and 35% of CHP plant is fuelled by waste and biomass.

**Storing peak energy** – Unlike electricity, hot water can be stored easily and cheaply. Excess electricity from wind generation can be converted into stored heat when electricity demand is low.

Responding to the '70s oil crises in the UK, Lord Marshall's 1979 energy paper recommended the adoption of CHP and district heating for a number of key cities. But these initiatives were

| Planning policy & regeneration | Developers required to:<br>• connect to any existing network<br>• consider site-wide networks on schemes<br>• be 'network ready' e.g. provide communal wet heating system. | • Only covers new build projects<br>• Connections always subject to viability<br>• For 3rd party heat network developers, planning (Section 106) agreements are not 'bankable' revenues – e.g. heat supply contract not guaranteed. |
| --- | --- | --- |
| Providing 'anchor' heat loads | • Offer long term agreements to purchase heat for own buildings and social housing. | • Asset disposal (most local authority portfolios are shrinking)<br>• Transfer and outsourcing (e.g. 3rd parties responsible for energy use – e.g. Academy schools, PFI leisure centres etc). |

**Table 1:** Tools used by local authorities.

soon lost in the flow of cheap gas from the North Sea and in gearing up the energy industry for privatisation. It has taken over thirty years for district heating to return to UK policy. 'The Future of Heating' (Department of Energy & Climate Change, 2013) includes a chapter dedicated to the role of heat networks in decarbonising heat in urban areas. The policy details for delivering these networks are expected for consultation in 2014.

### How can we make it work this time?

Local government has been ahead of national government on district heating, particularly in London, where successive Mayoral administrations have strongly supported the development of heat networks. The main tools used by local authorities are set out in **Table 1**, along with some of the constraints. But the biggest threat to further action is that all this activity is optional, at a time when local government has to decide what statutory services to cut back. So there needs to be a strong national framework of support to turn feasibility studies (of which we have plenty) into real networks.

The big carbon prize is in existing buildings, but here there are no incentives or policies to drive the creation of heat networks – and they will also be competing against well established electricity and gas networks with secure regulated revenues for maintenance and replacement.

### Fixing the heat off-take risk

Heat networks are capital intensive new infrastructure. In the absence of any incentives to create these networks, new district heating projects have to finance the whole system from generation plant, network and building heat exchangers, from heat and electricity revenues. Hence, most new schemes:

- are based on gas fuelled CHP as the lowest cost form of generation
- require either long-term guaranteed heat revenues (from existing building schemes), or one off connection fees to cover capital investment (in new build development schemes).

## Supporting the right infrastructure in the right place

Heat networks will only be the most cost-effective solution for low carbon heat in areas with sufficient heat density. But equally, we should not be incentivising ASHP or gas micro-CHP in those same areas as this will result in higher pass-through distribution costs to customers. The worst case will be where an upgraded electricity cable, gas pipe and heat main all pass down a street but each only supply heat to a third of the buildings.

The government's emerging policies will need to address these issues, but a good start would be:

- Local authorities, together with central government, undertake heat planning and agree to designate zones where heat networks will be pursued.
- CHP/district heating receives incentives within zones identified for heat networks – individual low-carbon installations do not (e.g. heat pumps, solar, biomass, micro-CHP).
- Buildings occupied by publically funded organisations must connect to district heating where it is demonstrated to be economically viable (to compensate for the fragmenting local authority estate).
- Consider a small levy on electricity and gas networks to be used to underwrite new heat networks (the levy being based on the avoided costs of upgrading these networks to carry greater capacity for heat).

## Ensuring new gas generation is CHP

It is now almost inevitable that new gas plant will be built to fill the gap between the closure of old plant and the delays in getting the new low carbon plant developed. As policy makers have viewed power stations solely from the point of view of electricity generation, use of the waste heat for district heating has not been properly rewarded because (as highlighted above) this decreases the electricity output and increases the carbon content of the electricity. Hence, this zero carbon district heat has never been on a level playing field with high carbon domestic gas boiler heat (which attracts no carbon taxation at all).

It is therefore encouraging that the government's *Future of Heating* document (Department of Energy & Climate Change, 2013) promises both to develop a new bespoke policy for CHP, and to treat our energy supply systems as an inter-related whole. But we also need to ensure that new plant is developed in proximity to heat demand (this will also reduce the electricity losses). We moved our power stations out of cities because of smog, we now need to bring them back – albeit with cleaner fuels, and technology to clean up the smoke stacks.

While thinking about London's energy future it's worth reminding ourselves what it was like fifty years ago. In 1962, Battersea power station was producing power for London while also operating as a CHP plant providing heat to London's first district heating scheme in Pimlico. Although fuelled by coal, the heat would have had a carbon content just below a condensing gas boiler. It was a great idea then, and it's still a great idea for the future.

## References

Department of Energy & Climate Change. 2013 (March). The Future of Heating: Meeting the Challenge. Available from: https://www.gov.uk/government/uploads/system/uploads/attachment_data/file/190149/16_04-DECC-The_Future_of_Heating_Accessible-10.pdf. [Accessed 7 August 2013]

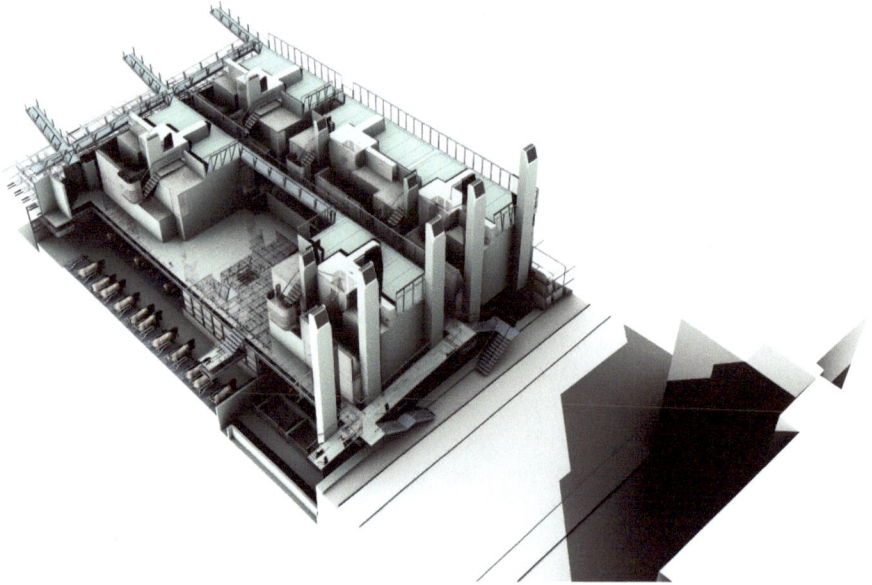

# Future-proofing London

## Sofie Pelsmakers

It is an undeniable fact that even a relatively minimal rise in global warming (around 0.8–1°C to date) has already had a discernible effect on our environment, causing sea level rises of up to 220mm (Allen et al, 2013; UKCP09, n.d). While the EU and UK governments have agreed that global warming must be limited to a maximum 2°C rise, global carbon dioxide ($CO_2$) emissions continue to rise at an alarming rate (Allen et al, 2013; UKCP09, n.d). Current records suggest that the world is heading towards a 2–6°C rise by 2100, associated with a medium, or even high risk global warming scenario (Jennings & Hulme, 2010). So what does this mean for our cities, and specifically for London?

In recent years, successive UK governments have made a commitment to ensuring that all new build dwellings are designed to 'zero carbon' by 2016, an objective to be rolled out across other building types by 2019 (DCLG, 2013). The government also envisages an upgrade of all existing buildings to zero, or close to 'zero carbon' by 2050 (DECC, 2011). This would suggest, if targets were met, that by 2062 the entire built environment in London will be virtually zero carbon.

Zero carbon buildings can only be achieved by building or upgrading existing structures to significantly increased building fabric energy efficiency standards, while also implementing on site renewable technologies and relying on a cleaner energy supply. In addition, it will also be necessary to make considerations for the projected impact of climate change on our existing and future designs, for it must be assumed that by 2062, our buildings and their occupants will be subject to different environmental conditions. Many UK buildings are designed with a lifespan of at least sixty years and 75% of existing UK housing is expected to still be in use in 2050 (SDC, 2006), so it is important that adaptation measures are considered now in order to withstand the climate changes predicted during that period.

### London's future predicted climate

As evidenced by the facts below, climate change is already a reality in the UK:

**How to cite this book chapter:**
Pelsmakers, S. 2013. Future-proofing London. In: Bell, S and Paskins, J. (eds.) *Imagining the Future City: London 2062.* Pp. 73-83. London: Ubiquity Press. DOI: http://dx.doi.org/10.5334/bag.k

- The UK has seen a 1°C temperature increase since 1970 (UKCP09, n.d).
- In the last 15 years London experienced the four hottest years on record, with 2006 exhibiting the highest temperatures in 350 years (UKCP09, n.d).
- When a prolonged heat wave hit the country in 2003, England and Wales suffered 2000 heat-related deaths (an increase of 16%) (UKCP09, n.d), with the greatest impact in London (Beizaee, 2013).

The discernible drop in temperature during the past few winters in the UK can be considered as further evidence of climate change. Global warming has caused record thawing of arctic sea ice, which in turn impacts on air and wind patterns worldwide. Instead of the milder Atlantic conditions usually experienced, atmospheric changes have caused a flow of cold Arctic weather to reach the UK (Ritter, 2013). It is likely that London will continue to see similarly cold and variable winters over the next few years, as a result of climate change induced ice melting (Huffington Post, 2013).

Long-term however, London's year round climate is predicted to rise by 2062, with warmer, drier summers, milder, wetter winters, and more extreme winds and rainfall (UKCP09, n.d). While increases in temperature will be incremental, the actual impact on the natural and built environment is expected to be significant.

It is likely that London will experience more extreme weather conditions, such as flooding, heat waves and droughts, the latter of which may increase the incidence of subsidence (Shaw et al, 2007). Although the level of annual rainfall is not predicted to vary significantly, it is the distribution of rain which will become problematic, with more rain falling in winter (possibly up to a third more by 2062) and nearly equivalent decreases in summer. Temperatures in London in 2062 are also expected to rise by around 3–4°C in the summer and 2.5–3°C in winter (UKCP09, n.d). As such, $CO_2$ emissions from space heating might decrease, but this is unlikely to cancel out the predicted increases in cooling energy during the summer.

Indeed, $CO_2$ emissions from summer cooling are likely to increase significantly, exacerbating global warming, local air pollution and the Urban Heat Island effect in cities. Within these conditions, ill adapted buildings will struggle to provide the necessary thermal comfort, leading to heat-related deaths and illness, especially among the elderly and vulnerable. Prolonged temperatures over 35°C could also cause road surfaces to melt and other urban infrastructures to fail (UKCP09, n.d). Figure 1 illustrates this impact, where summer air temperatures in London might resemble Marseille's by 2080, but without the benefit of extra sun hours in winter.

Although these predicted weather conditions will be new to London, similar climatic conditions can be witnessed elsewhere around the world, and as such, provide us with existing strategies to prepare accordingly. Mediterranean cities can teach London how to adapt buildings to cope with increased summer temperatures, while cities in the Netherlands, a country with more than 50% of its landmass below sea level, demonstrate how it might be possible to work *with* water, rather than against it to prevent flooding.

So what is the best way to combine mitigation strategies, while simultaneously building in adaptation measures to ensure our built environment continues to perform in a changing climate?

## Adapting our cities for the future

The fact that we are operating in a changing climate can no longer be ignored. If London's buildings and infrastructure are not designed or adapted to cope, then temperature related health issues and flood damage will become significant problems (DEFRA, 2012). As set out below, it is critical

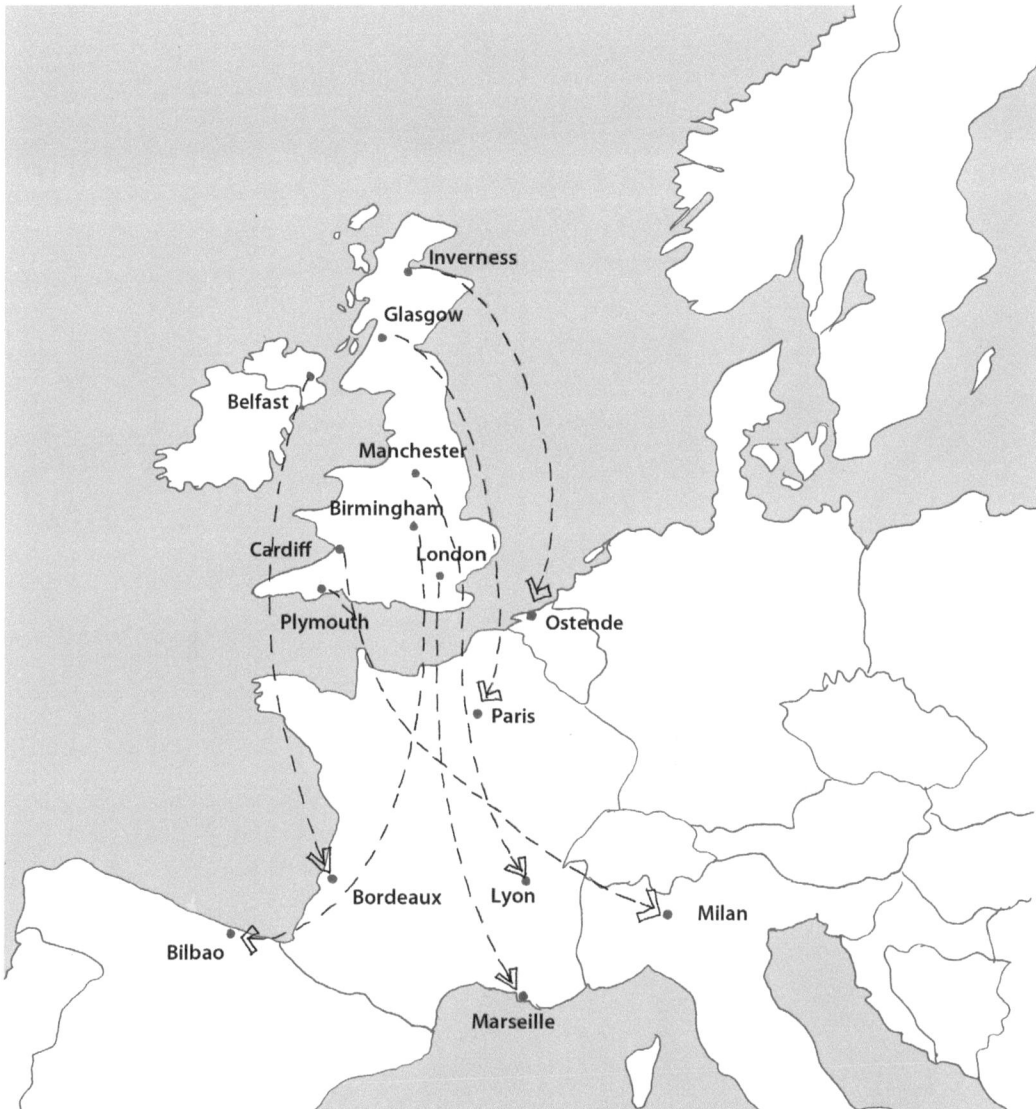

**Figure 1:** Predicted summer air temperatures in UK cities by 2080 (Pelsmakers, 2012).

that adaptation strategies are implemented early to prevent buildings from overheating and to safeguard occupants and property from floodwaters.

### *Insulation and airtightness*

High fabric energy efficiency in the form of increased insulation and airtightness are key measures in the current climate for a sustainable built future, generally resulting in reduced space heat demand, increased thermal comfort and reduced fuel poverty. Despite the long-term prediction of increasingly mild winters, even super-insulated buildings will continue to require some winter space heating. In addition to helping to retain this heat, good insulation also buffers the internal environment from the external elements, whether cold or hot. So insulating in a warming

## Most effective shading

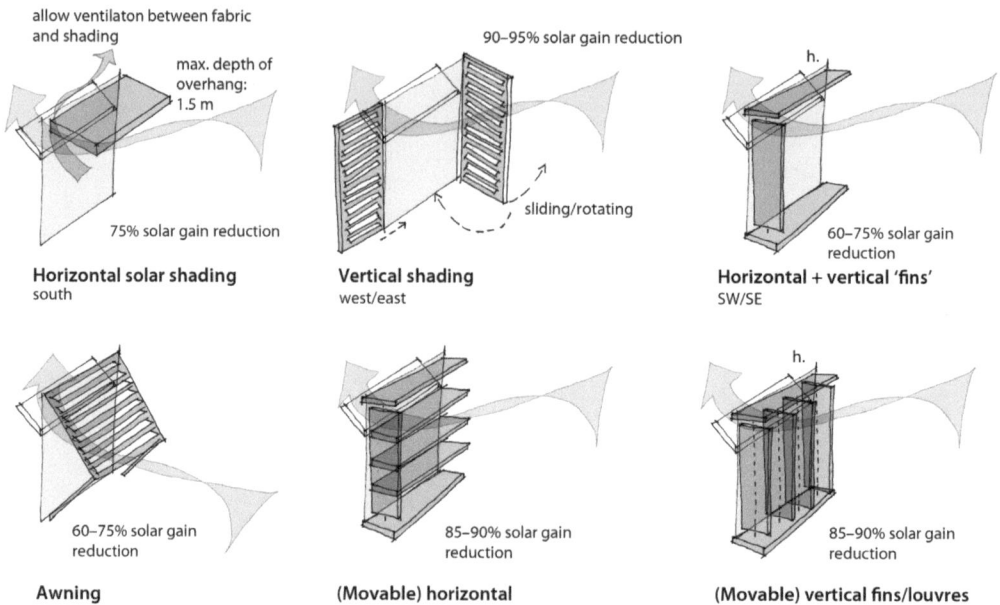

allow ventilaton between fabric
and shading

max. depth of
overhang:
1.5 m

90–95% solar gain reduction

h.

75% solar gain reduction

sliding/rotating

60–75% solar gain
reduction

**Horizontal solar shading**
south

**Vertical shading**
west/east

**Horizontal + vertical 'fins'**
SW/SE

60–75% solar gain
reduction

85–90% solar gain
reduction

h.

85–90% solar gain
reduction

**Awning**

**(Movable) horizontal**

**(Movable) vertical fins/louvres**

Bottom three shading devices are suitable for all orientations if movable shading fins.
They are effective solar shading, but reduce daylighting and winter solar gain so use with care.

Design sliding/inward-opening windows, which do not impede natural ventilation. Design top
inward-opening 'hopper' windows for night cooling (h.).

**Figure 2:** Effective solar shading devices dependant on orientation, while allowing for natural ventilation with inward opening windows. The illustrated top 'hopper' window is ideal for secure night cooling in thermally massive buildings (Pelsmakers, 2012).

climate generally still makes sense, and provides future-proofing according to short-term and long-term scenarios.

### *Overheating prevention: building scale*

The prevention of overheating is important at both citywide and building levels, particularly as increased heat-related deaths occur even in relatively low external temperatures of around 19°C (Beizaee et al, 2013). However, it is not the increased temperatures per se that are dangerous, after all, there are many countries with far hotter climates. It is the irregularities in temperature, and the intensity and duration that people in the UK are unaccustomed to (Kalkstein, 2000; Vandentorren et al, 2006). Furthermore, UK buildings are often so ill adapted that indoor air temperatures can actually exceed those externally. This generates dangerous living conditions for those at risk, as internal temperatures above 35°C increase cardiovascular stress and the danger of respiratory diseases, especially in areas where local air pollution is high (Hacker et al, 2005).

The external temperature thresholds defined by the MetOffice for severe hot weather warnings are outlined as 32°C in the daytime and 18°C at night (Gething & Puckett, 2013). Research has shown that even in mild summers recommended summer comfort temperatures and overheating thresholds (26°C in bedrooms; 28°C in living rooms, offices and schools (CIBSE 2005; 2006)) are already being exceeded by more than 1% of occupied houses studied in London and the South East, particularly in newer constructions (Beizaee et al, 2013). To prevent such risks, measures need to be incorporated into new building designs *now*, to ensure that internal temperatures can be maintained at a healthy, comfortable level for occupants, today and in the future.

Overheating is not caused by super-insulating buildings, but usually through a combination of high internal heat gains, a lack of summer solar shading, and ill designed or absent night ventilation in well insulated buildings. Incorporating external solar shading, combined with sliding or inward opening windows to allow good natural ventilation (as illustrated in Figure 2) are crucial adaptation strategies in both new build and refurbished building designs to prevent overheating.

In some cases however, the addition of external shading can cause significant structural difficulties when retrofitting, so planners may encourage less effective internal shading methods instead. Other good low energy practices, such as the specification of energy efficient appliances and the reduction of artificial lighting through good day-lighting design, can also help reduce unwanted internal summer heat gains. Additionally, as in Mediterranean traditions, taller floor to ceiling heights (greater than 2.7 metres) are useful for allowing the natural stratification of hot summer air to rise well above head height.

The presence of thermal mass in buildings can provide summer cooling of 3–5°C, generating significant decreases in summer cooling energy demand (Hacker et al, 2008). This is only possible however, if secure night cooling is implemented to release built-up heat from the daytime, as failure to do so may cause overheating. When refurbishing, covering up solid walls with insulation can significantly decrease the available thermal mass of existing buildings and therefore, internal wall insulation upgrades need to be carefully considered to ensure continued summer thermal comfort in a warming climate.

If buildings are ill adapted to extreme temperatures, occupants may resort to the ad hoc installation of air conditioning units, generating significant increases in a building's typical operational energy use (Hacker et al, 2005). Air conditioning units also operate by 'dumping' excess heat outside the building, which only serves to augment external temperatures and exacerbate the Urban Heat Island effect (Tremeac et al, 2012; Vardoulakis & Heaviside, 2012), and should therefore be avoided. Instead, future buildings must be designed to promote lower energy consumption through the use of natural ventilation and 'adaptive comfort' measures, such as the occupants' ability to adapt their clothing and manipulate windows and shutters for thermal comfort. While internal temperatures will never be as low as actively air conditioned buildings, designing a naturally managed building is about tempering the external environment within acceptable limits based on occupant control, which has been found to increase occupant satisfaction. It is the perceived difference between internal and external conditions that has been found to be more important than the actual temperatures achieved. In fact, air conditioned buildings have been associated with decreased occupant wellbeing (Steemers & Manchanda, 2010).

### Overheating prevention: London scale

Due to the Urban Heat Island effect, the impact of climate change is noticeably amplified within cities. The Urban Heat Island effect is generated by a combination of human activity, a lack of green spaces, plus the dark, thermally massive surfaces which constitute our cities. This results in an approximate rise in temperature of around 4–5°C in urban areas compared to the surrounding countryside.

At a city scale, heat build up can be prevented by introducing more green spaces throughout urban areas, particularly those that include 'water squares'. The larger the green space, the greater its tempering effect. This effect can be felt within spaces as small as 10m diameter and on average, park areas tend to remain 2–3°C cooler than surrounding streets (Graves et al, 2001). Vegetation also provides some relief from the sun, reducing the elevated risk of skin cancer caused by prolonged exposure outdoors (DEFRA, 2012; Vardoulakis & Heaviside, 2012).

Approximate summer surface temperatures °C in brackets

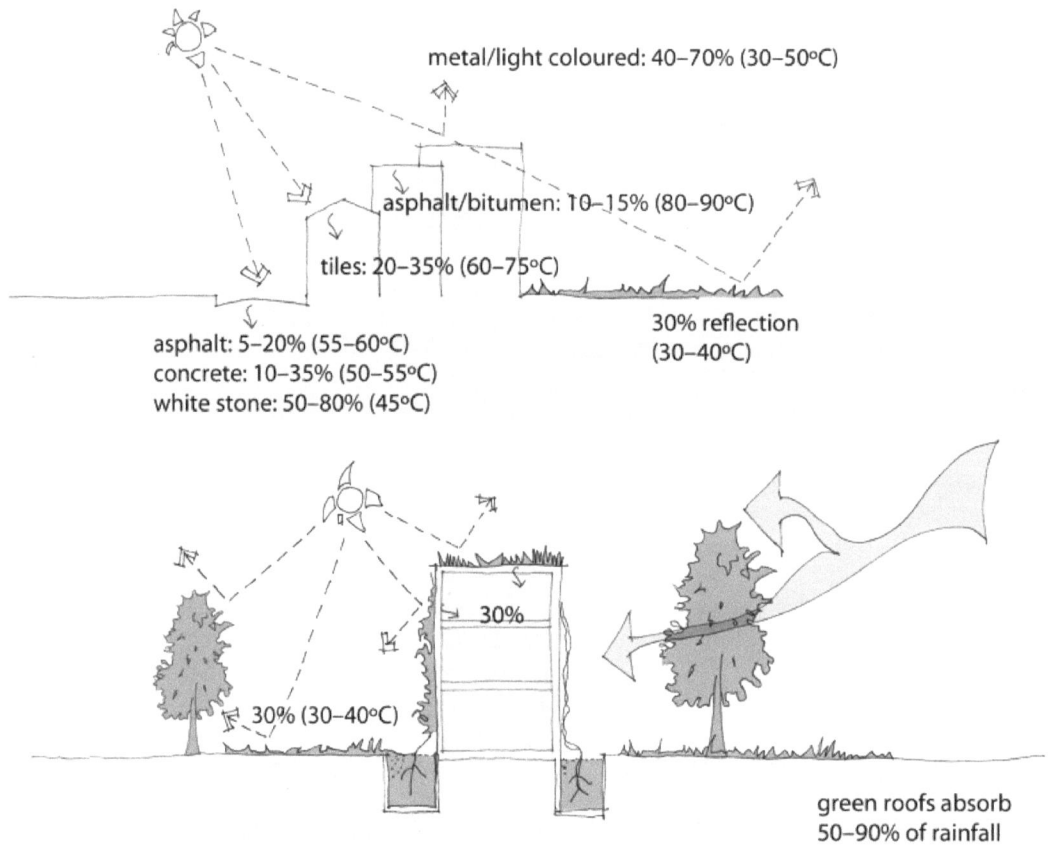

metal/light coloured: 40–70% (30–50°C)

asphalt/bitumen: 10–15% (80–90°C)

tiles: 20–35% (60–75°C)

30% reflection
(30–40°C)

asphalt: 5–20% (55–60°C)
concrete: 10–35% (50–55°C)
white stone: 50–80% (45°C)

30%

30% (30–40°C)

green roofs absorb
50–90% of rainfall

**Figure 3:** Typical solar reflection and surface temperatures of different materials, with a typical city's material built up at the top and a resilient city's built up below (Pelsmakers, 2012).

Implementing light, reflective external surfaces helps generate an 'urban cooling' effect by reducing the surface temperature of a building or infrastructure by around 10–20°C (Erell et al, 2011; Santamouris & Asimakopoulos, 2001; Ward, 2004), as illustrated in Figure 3. An example of such adaptive measures can be seen in the specification of road surfaces in the UK, which have already been modified to ensure the materials can withstand greater temperatures and allow for increased drainage capacity following feedback from previous heat waves and flood events (DEFRA, 2012).

*Flood prevention: London scale*

There are around 2.5 million dwellings at risk of flooding in England and Wales (Carrington & Salvidge, 2007; DCLG, n.d). In England, 11% of dwellings are built in flood risk areas, most of which (around 850,000 properties) are in London. This means that up to 25% of homes in London are in danger, the majority of which at risk of river-flood damage, especially in the boroughs of Southwark, Hammersmith and Fulham, Wandsworth and Newham (Carrington & Salvidge, 2007). Richmond Upon Thames holds the largest number of properties in the 'highest risk' category, but all London boroughs are at an increased risk of flash floods, due to the abundance of hard surfaces in the city combined with increased downpours and low capacity urban drainage systems (Carrington & Salvidge, 2007).

Projected changes in climate are only likely to increase the risk of flooding, as global warming will continue to alter the intensity, duration and distribution of rainfall, resulting in more urban, and flash, flooding (ABI, n.d(a); n.d(b)). Furthermore, while risks are increasing, flood defence cutbacks threaten to make home insurance unaffordable for many (ABI, 2013; Carrington & Salvidge, 2007).

Flooding is not only disruptive at an operational level, causing problems for the London Underground system and ground floor infrastructure, sewage and clean water supply of the capital, but it has health implications for the public (Vardoulakis & Heaviside, 2012). Water as shallow as 15cm deep can be threatening to those at risk, particularly the elderly or young, with mortality rates highest in flash floods (Vardoulakis & Heaviside, 2012).

In order to future-proof our cities against these challenges, it is important to learn how to work *with* water, rather than against it. This includes undertaking flood risk assessments, allowing a minimum of 5% space for water storage on a site and the provision of efficient water flow channels (DCLG, 2006). Furthermore, green spaces, water squares and increased permeable surfaces can also help, for they not only reduce local summer air temperatures, but collect rainwater and aid water run off. This can reduce localised flash flood risk and acts as an amenity for city dwellers.

### *Flood prevention: building scale*

To prevent the threatening consequences of flooding, such as loss of property, injury, disease and even death, it is vital that building adaptations are designed as an integral element of future city-wide protections (DEFRA, 2012). At a building scale, considerations must be made for zoning, structural adaptations, and even the use of different typologies (illustrated in Figure 4), such as sacrificial ground floors, buildings on stilts, or floating buildings. Most existing structures can be 'wet-proofed', which means they are designed with possible future flooding in mind and result in only minimal damage to the property should this happen. This can be achieved through the use of water resistant materials for floors, walls and fixtures, and the siting of electrical controls, cables and appliances at greater than one metre height (DCLG, 2007).

### Climate change adaptation is a necessity, not a luxury

Whatever the future holds, we cannot afford to be complacent – especially when considering these measures and many more can easily be incorporated into current design procedures as part of good new build and refurbishment practice.

Some measures can be implemented later, while others need to be effected immediately. For instance with new build, careful site planning and consideration for the orientation of a building must happen now, as with the specification of a well insulated building fabric and inward opening windows. However, intensive green or brown roofs could be implemented in the future, to reduce water run off and aid 'urban cooling' so long as the additional structural impact is taken into account and fixings are incorporated into the initial designs to make future adaptations easy and feasible.

**A. Sacrificial basement/ ground floor**

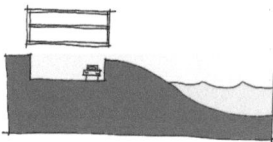

A raised ground floor with water retention in the basement (storage/car park) area, with residential at first floor. Suitable in low to medium flood probability zones.

This typology can lead to poor-quality street levels and security issues. Better to put workshop uses with less vulnerability on ground floor and vulnerable uses above to create active frontages. Move all equipment etc. above flood risk level.

**B. Building on stilts**

Useful in a flood inundation area; but still needs protection from breakwaters to avoid debris damaging stilts structurally. Suitable in high flood probability zones. Difficult in an urban environment: issue with the undercroft and aerial walkways, usually lacking surveillance and ownership, leading to poor-quality street level and security issues. Lift or ramps required. Nothing can really grow under the stilts, to utilise as open space. Workshop/community infill space could be useful.

**C. Floating buildings**

Ground floor rises with water levels up to around 5.5 m. Usually built with EPS polystyrene slabs, with concrete screed over, to achieve floating ground floor base. Suitable in high flood probability zones. Good for areas where inundation can be controlled. Building on water avoids the need to reclaim land. Services to be encased in flexible pipework to allow vertical movement of buildings. Connection to a floating pier and mooring posts required. Most suited to smaller buildings. Usually low thermal mass buildings, which may lead to issues of summer overheating. Not many UK precedents, but more common in Netherlands.

**D. Flood-resilient or 'wet-proofed' buildings**

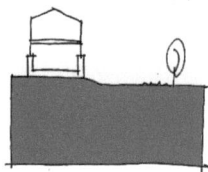

'Wet-proofed' buildings are designed with possible future flooding in mind and with minimal damage to the property when this happens. This may be achieved through the use of water-resistant materials for floors, walls and fixtures and the siting of electrical controls, cables and appliances at a higher than normal level. If the lowest floor level is raised above the predicted flood level, allow ramp access for disabled users. Suitable in all flood probability zones.

**Figure 4:** Flood mitigation: building typologies that work with water (Pelsmakers, 2012).

Neglecting to future-proof our buildings will only result in a city ill adapted to the future needs of our society within a changing local and global environment. Buildings will fail to function effectively under extreme weather conditions leading to increased, wasteful energy use, and exacerbating the effects of global warming. At worst, the inability of our built environment to cater to the demands of its inhabitants might simply result in a stock of obsolete, unhealthy buildings unfit for purpose.

Designing for climate change adaptation on the other hand is guaranteed to increase building lifespan and help protect occupants from the detrimental effects of global warming, while also reducing the necessity for costly and carbon intensive interventions in years to come. Furthermore, there are other beneficial side effects to many adaptation measures, such as increased occupant satisfaction and wellbeing, and the support of London's declining wildlife through increased green spaces and trees (RSPB, 2013).

The effects of climate change are an undeniable reality and to safeguard our cities from extensive damage we need to start designing for these changes right now. With sufficient foresight and planning, we can provide buildings which aid mitigation efforts and, when needed in the future, also support the ongoing adaptation of our cities for years to come.

# References

ABI. 2013 (30 May). Insures continue to offer households access to insurance while negotiations with the Government continue. Available from: https://www.abi.org.uk/News/News-releases/2013/05/Insurers-continue-to-offer-householders-access-to-flood-insurance. [Accessed 17 August 2013]

ABI. n.d(a). Climate Adaptation Guidance on Insurance Issues for New Developments. Association of British Insurers. Available from: https://www.abi.org.uk/~/media/Files/Documents/Publications/Public/Migrated/Flooding/Climate%20adaption%20guidance%20on%20insurance%20issues%20for%20new%20developments.ashx. [Accessed 17 August 2013]

ABI. n.d(b). Rising global temperatures will put the heat on insurance as flood costs rise new research from the ABI. Available from: https://www.abi.org.uk/News/News-releases/2009/11/Rising-global-temperatures-will-put-the-heat-on-insurance-as-flood-costs-rise--new-research-from-the-ABI. [Accessed 17 August 2013]

Allen P., Blake L., Harper P., Hooker-Stroud A., James P. and Kellner T. 2013. *Zero Carbon Britain.* Machynlleth: Centre for Alternative Technology

Beizaee A., Lomas K.J, Firth, S.K. 2013. National survey of summertime temperatures and overheating risk in English homes. *Building and Environment.* 65:1-17.

Carrington D. and Salvidge R. 2013 (17 May) Flooding threatens one in four London properties. The Guardian. Available from: http://www.guardian.co.uk/environment/2013/may/17/flooding-threat-london-property. [Accessed 17 August 2013]

CIBSE. 2005. TM36 *Climate change and the indoor environment: impacts and adaptation.* London: CIBSE

CIBSE. 2006. *Guide A Environmental design.* London: CIBSE

DCLG. 2006. *Planning Policy Statement 25: Development and Flood Risk.* London: The Stationery Office

DCLG. 2007. *Improving the flood performance of new buildings.* London: Department For Communities and Local Government

DCLG. 2013. Improving the energy efficiency of buildings and using planning to protect the environment. Available from: https://www.gov.uk/government/policies/improving-the-energy-efficiency-of-buildings-and-using-planning-to-protect-the-environment. [Accessed 17 August 2013]

DCLG. n.d. Live tables on land use change statistics. Available from: http://www.communities.gov.uk/planningandbuilding/planningbuilding/planningstatistics/livetables/landusechange/. [Accessed 17 August 2013]

DECC. 2011. *The Carbon Plan: Delivering our low carbon future.* London: Department of Energy and Climate Change

DEFRA. 2012. *UK climate change risk-assessment: government report.* London: The Stationary Office

Erell E., Pearlmutter D. and Williamson T. 2011. *Urban Microclimate: Designing the Spaces Between Buildings.* London: Earthscan

Gething B. and Puckett K. 2013. *Design for Climate Change.* London: RIBA Publishing

Graves H., Watkins R., Westbury P. and Littlefair P. 2001. *Cooling buildings in London: overcoming the heat island.* London: BRE

Hacker J.N., Belcher S.E. and Connell R.K. 2005. *Beating the Heat: keeping UK buildings cool in a warming climate. UKCIP Briefing Report.* Oxford: UKCIP

Hacker J., De Saulles T.P., Minson A.J. and Holmes M. 2008. Embodied and operational carbon emissions from housing: a case study on the effects of thermal mass and climate change. *Energy and Buildings.* 40: 375-384.

Huffington Post. 2013 (11 April). Climate Change 'Causing Colder British Winters' Says Met Office Chief Scientist. *The Huffington Post, United Kingdom.* Available from: http://www.huffingtonpost.co.uk/2013/04/11/climate-change-colder-winter-met-office-chief-scientist-_n_3059116.html. [Accessed 17 August 2013]

Jennings N. and Hulme M. 2010. UK newspaper (mis)representations of the potential for a collapse of the Thermohaline Circulation. *Area.* 42(4): 222-456.

Kalkstein L. 2000. Saving Lives during Extreme Summer Weather. *British Medical Journal.* 321: 650.

Pelsmakers S. 2012. *The Environmental Design Pocketbook.* London: RIBA Publishing

Ritter K. 2013 (29 March). Q&A Europe's freezing Easter and global warming. Available from: http://phys.org/news/2013-03-qa-europe-easter-global.html. [Accessed 17 August 2013]

RSPB. 2013. State of Nature 2013. [Accessed 17 August 2013]; Available from: http://www.rspb.org.uk/stateofnature.

Santamouris M. and Asimakopoulos. 2001. *Energy and Climate in the Urban Built Environment.* London: Earthscan

SDC. 2006. Stock Take: Delivering Improvements in Existing Housing. London: Sustainable Development Commission

Shaw R., Colley M. and Connell R. 2007. *Climate Change Adaptation by Design.* Oldham: Town and Country Planning Association

Steemers K. and Manchanda S. 2010. Energy efficient design and occupant well-being: Case studies in the UK and India. *Building and Environment.* 45(2): 270-278.

Tremeac B., Bousquet P., de Munck C., Pigeon G., Masson V., Marchadier C., Merchat M., Poeuf P. and Meunier F. 2012. Influence of air conditioning management on heat island in Paris air street temperatures. *Applied Energy.* 95: 102-110.

UKCP09. n.d. UK Climate Projections. Available from: http://ukclimateprojections.defra.gov.uk/. [Accessed 17 August 2013]

Vandentorren S., Bretin P., Mandereau-Bruno L., Crosier A., Cochet C., Riberon J., Siberan I., Declercq B. and Ledrans M. 2006. August 2003 Heat wave in France: risk factors for death of elderly people living at home. *European Journal of Public Health.* 16(6): 583-591.

Vardoulakis S. and Heaviside C. 2012. *Health Effects of Climate Change in the UK 2012.* London: Health Protection Agency

Ward I. 2004. *Energy and Environmental Issues for the Practising Architect.* London: Thomas Telford

# Water supply, drainage and flood protection

## Sarah Bell

Water, drainage, sewerage and flood defence infrastructure are vital for the success of any city. These systems also provide insights into how cities relate to their natural environments. Water provides a direct connection between people and landscapes in everyday urban life. The water we splash on our faces or flush down our toilets first thing in the morning provides a real, tangible connection with both the hidden landscape of urban infrastructure and the hydrological landscape that extends beyond the city. Considering water in London in 2062 is therefore more than simply a question of supply, demand, rainfall and flood risk. Our future water systems will embody our assumptions about, and relationship to, the landscapes we are part of.

Our hydrological landscapes are changing. The UK Climate Impacts Programme 2009 projections forecast that by the 2050s London's average rainfall is unlikely to change, with the most likely average rainfall between 5% more or 5% less than now. By the 2080s the central forecast remains for no change in average rainfall, within a range of 7% more or 6% less than now. However, the timing of rainfall throughout the year is likely to change by the second half of the century, with wetter winters and drier summers (UKCP09, n.d.). This has significant implications for a water supply and drainage systems designed for relatively stable and consistent monthly rainfall. Changes in rainfall will increase flood risk, but more significant to flooding in London are forecast sea level rises of between 22.2cm and 31.4cm by 2060, when our current flood defence systems will be reaching the end of their effective life spans (Environment Agency, 2012; UKCP09, n.d).

This chapter addresses three key elements of water infrastructure in London in 2062: water supply and demand, surface water and flooding. Current systems and plans from infrastructure providers and government agencies are reviewed, and alternatives explored. The chapter considers some of the forecasts for water itself, how much there will be and how we will interact with it, the institutional arrangements for infrastructure, and how the challenges of dealing with water in London over the next fifty years might provoke radical change in how London relates to its natural environment.

**How to cite this book chapter:**
Bell, S. 2013. Water supply, drainage and flood protection. In: Bell, S and Paskins, J. (eds.) *Imagining the Future City: London 2062*. Pp. 85-93. London: Ubiquity Press. DOI: http://dx.doi.org/10.5334/bag.l

## Water supply and demand

London relies on river flows for its water supply. More than 70% of water used in London is abstracted from the River Thames, upstream of Teddington Weir (Thames Water, n.d). The remainder comes from other rivers, including the Lee, and groundwater. A desalination plant at Beckton is intended to be used at full capacity only during drought periods. Higher variability in annual rainfall is likely to impact on surface water flows, providing challenges for how water resources are managed for London throughout the year. The possibility of more frequent and intense droughts, resulting from greater variability in inter-annual rainfall presents further complexity for London's water managers and planners to deal with (UKCP09, n.d.).

A bigger challenge for London's water managers over the next fifty years will come from the demands of a growing population. London's water and sewerage services are provided by privatised water companies, the largest of which is Thames Water, covering most of the city. Thames Water's 2015-2040 Water Resources Management Plan forecasts a deficit between current supply capacity and future demand in London of 367 megalitres per day, or the equivalent supply to 2.2 million household customers by 2040 (Thames Water, n.d). This is consistent with forecasts for a population of between 9-10 million people by the middle of the century, approximately 2 million more than now (Greater London Authority, 2012).

The Thames Basin and much of the South East of England are currently classified as areas of serious water stress by the Environment Agency, without taking account of potential future impacts of climate change (Environment Agency. 2008). We currently take more water from the environment than it can sustain, with detrimental impacts on aquatic ecosystems and rural landscapes. There is no more water to be abstracted from the environment to supply a growing population in London. We cannot drill more boreholes, or pump more water from our rivers, without irreparably damaging our landscapes and ecosystems.

Butting up against the hydrological limits of our landscape presents us with the opportunity to rethink how we live with water in London. Thames Water is obliged to meet all demands for water from its customers, and so they must plan conservatively. Their plans for addressing the projected shortfall between supply and demand are based on reducing leakage in their networks, reducing per capita demand from customers, increasing transfers of water from other catchments and water recycling. Thames Water is currently the sole provider of water infrastructure services in London, but responsibility for water must be shared more broadly. By 2062 the structure of the water industry, and our concepts of water infrastructure and services, may have changed significantly.

### Demand Management

Reducing the per capita demand for water is the lowest cost solution to addressing the projected supply shortfall. Londoners currently use more water on average that the rest of the UK: 167 compared to 150 litres per person per day (Nickson et al, 2011). The UK government has a target to reduce the national average to 130 litres per person per day by 2030, through a combination of water metering and water efficiency measures (DEFRA, 2011). Thames Water's projections for savings due to water demand measures are more conservative than government targets, and assume their customers will be using 141 litres per person per day by 2040 (Thames Water, n.d). Whilst Thames Water is understandably conservative in their estimates of the impact of demand management, it is possible that by 2062 average water demand in London could be as low as 100 litres per person per day, based on design estimates of 80 litres per person per day for zero carbon homes (DCLG, 2010).

Currently only 30% of Thames Water's customers are metered, a figure that they plan to increase to 78% by 2040 (Thames Water, n.d). Further increases in meter penetration may be possible by

2062, particularly as hard to meter properties such as flats are demolished and reconstructed, but full penetration is unlikely considering the long life of London's building stock and plumbing systems. Thames Water estimate that metering will reduce demand by fifty megalitres per day by 2025, due to better leak detection and customer water conservation (Thames Water, n.d).

Most new water metering will be 'smart' meters. By 2062 it is likely that smart metering and sensor technologies will be considerably cheaper, allowing for much wider application. Not only does smart metering provide additional feedback to users, but it also allows for more specificity in tariffs. Ubiquitous sensing and smart metering applied to domestic water use in London 2062 could mean that customers pay more for water during a peak demand periods, or that they pay more for water used in the garden than the kitchen. Ensuring that tariffs don't unfairly disadvantage vulnerable customers or those on low incomes will be an increasingly important function of the water industry regulator and customer representative groups.

Water efficiency measures involve behavioural and technological change (Butler & Memon, 2006). Implementing stronger water efficiency standards in building regulations and planning guidance will be a key driver for achieving high levels of water efficiency in new buildings. The Code for Sustainable Homes includes water efficiency standards, with 80 litres per person per day as the designed consumption in the highest Level 6 homes (DCLG, 2010). Retrofitting cistern displacement devices, flow-reducing valves, shower timers and other small 'gadgets' is a focus of water company efficiency campaigns. Over the next fifty years there will be greater opportunities for more substantial retrofitting for water efficiency. Even though most of London's building stock will remain, most bathrooms, kitchens and water using appliances will be replaced by 2062. Significant water efficiency improvements could be made by setting minimum standards for fittings and appliances available on the UK market, or providing subsidies or other incentives for the most efficient devices. Retrofitting rainwater harvesting and greywater reuse systems could also be encouraged by policy and economic measures, particularly in renovation or redevelopment projects requiring local authority planning consent.

### Alternative Water Systems

Currently in London, all domestic demand is met by potable supply from the Thames Water network, even though only a small proportion of water is used for drinking and cooking. Approximately 30% of water supplied is used to flush toilets, and a further 10% is used for laundry (Waterwise, n.d). These are examples of low risk water uses that could be supplied by clean water from alternative sources. Alternative supply systems provide water of sufficient quality for non-potable use, without requiring abstraction, treatment and distribution through the centralised water network. Rainwater harvesting, greywater reuse, surface water capture and other technologies provide potential sources of non-potable water in London.

Non-potable water supply systems at building or district scale in London could provide a considerable new source of water, outside the conventional infrastructure network. Rainwater harvesting or greywater reuse are relatively straightforward to implement in new buildings, but there is also considerable scope for retrofitting non-potable water supplies in London buildings, many of which have existing cisterns or header tanks that effectively supply a non-potable water system within the building, using water sourced from the potable network. Retrofitting rainwater or greywater collection, storage and filtration systems remains a considerable challenge, but the separation of potable mains supply and non-potable header tank supply in buildings reduced the need for completely re-plumbing buildings for non-potable use. The carbon emissions of rainwater harvesting systems are currently estimated to be higher than mains supply, but this is likely to have changed by 2062, as the carbon intensity of mains supply increases, and the relative efficiency of building scale systems improves (Parkes et al, 2010).

District scale reuse of relatively high quality wastewater is another option currently being investigated as a future source of water in London (Bell et al, 2013). Rather than treating municipal wastewater, which is contaminated with human faeces, this proposition involves redistributing water that has been used for relatively low risk purposes. Domestic scale reuse of this kind might involve using shower water to flush toilets, while at a district level this could involve water from a hair salon being redistributed locally for toilet flushing in a nearby pub, or cooling water from university laboratory equipment being used for landscape irrigation. Such systems would require relatively minor treatment, but would require retrofitting distribution networks between sources and points of demand for water. Although complex, opportunities for renewal of existing infrastructure or installation of new infrastructure such as district heating, indicate that installing non-potable water networks might be a feature of some parts of London in 2062.

Non-potable reuse of municipal wastewater is also a potential source of water for London in 2062, though it is unlikely to be widely implemented in the city. In 2012 Thames Water supplied non-potable water to the Olympic Park for toilet flushing and landscape irrigation (Knight et al, 2012). The system treats water abstracted from the northern outfall sewer, to non-potable standards, and distributes it using a dedicated network. Although the water is not treated to potable standards, the quality is much higher than required for its intended use to manage the risk of cross-connection by customers, leading to non-potable water being plumbed into potable systems such a drinking water taps. Treating sewage to this standard is energy intensive, but comparable with the total energy required for drinking water and wastewater treatment.

Alternative water systems do not easily fit within the current model of water infrastructure provision in London, which is focussed on the centralised utility company Thames Water. The UK is unique in the world in having a fully privatised water sector, and further competition in the sector may provide opportunities for a greater diversity of actors to enter the market for provision of water services over the next fifty years (DEFRA, 2011). Alternatively, greater government intervention to support alternative water systems would also open opportunities for a range of new organisations to provide new water services. For instance, wide scale implementation of rainwater harvesting systems provides a new market for servicing and maintaining these systems. Provision of non-potable water to homes and businesses could represent an entirely new infrastructure service, which could be provided be Thames Water or more specialist suppliers. Distributed water supplies for non-potable use will require new regulatory and design standards, and new structures for governance to manage public health risks and environmental impacts.

### New Supplies

Such possibilities for radical reduction in demand for potable water are, unsurprisingly, not included in Thames Water's Water Resources Management Plan. As a utility with a statutory obligation to provide customers with potable water and wastewater services, their plans for new supplies focus on options for potable water, supplied through the existing distribution network. Options include building a new reservoir to increase water storage, expanding desalination, buying water from neighbouring water companies and re-using municipal wastewater. The preferred option for new supply until 2040 is re-using wastewater.

Potable reuse of water involves treating the effluent from current wastewater treatment plants to a very high standard using reverse osmosis, then returning to raw water supplies, such as to the river immediately upstream from an abstraction point, a raw water reservoir, or to an aquifer used for water supply. The water is then treated again through conventional drinking water system and distributed in the potable supply network. Reverse osmosis is an energy intensive process, effectively removing molecular scale contaminants, and is the dominant technology currently used in desalination plants. The energy required, and therefore the cost, for treating wastewater effluent is considerably less than seawater, which is why water reuse is preferred to desalination in Thames Water's plan.

Public acceptability of potable reuse will be central to successful implementation (Bell & Aitken, 2006; Hartley, 2006). Proposals for similar projects in the US and Australia have failed because of public concern about drinking recycled sewage. Public acceptability of water reuse is a complex social, psychological and political problem, but research undertaken as part of Thames Water's feasibility study indicates that Londoner's are more likely to accept this new source of water than people in other places. Factors contributing to this include the urban myth that London drinking water has already been through seven sets of kidneys. Ensuring that public concerns are adequately addressed, including concerns about wider issues of water management such as leakage and demand management, will be essential if Thames Water are to build on early indications of acceptance and avoid considerable controversy.

The possibility of transfers of water to London from other catchments and water companies is explored, but dismissed as a significant new source of water for London in the Thames Water plan until 2040. Reforms in water markets in the coming decade could make water trading between companies easier, however the general conditions of water scarcity in the South East of England will constrain the amount of 'new water' this makes available to London. Increased water trading could result in agricultural or industrial users of water selling their resource allocations to Thames Water, particularly during drought years, but this is unlikely to considerably change the supply-demand balance in London.

Over the next fifty years it is possible that the idea of a 'national water grid' to move water from wetter parts of England and Wales to London will no longer be discussed in the media, and in engineering institutions, as it is unlikely that such infrastructure will be implemented without considerable investment and support from central government. More economically, ecologically and energetically favourable solutions are likely to be implemented in London, including radical demand management and water recycling, before a national scale water distribution network will be feasible. The national grid for water reflects nineteenth and twentieth century approaches to infrastructure, with visions of big engineering projects based on dams, pipes and tunnels. Such models of infrastructure will seem increasingly outdated in London in 2062, with smart city technologies helping to manage demand and providing high levels of control to improve the management of alternative water supply systems, and membrane technologies improving the efficiency of water recycling for potable reuse. Managing London's water resources within its own catchment will demonstrate a sound environmental ethic as well as being economically sensible. Schemes for pumping water across the country reflect nineteenth century determination to conquer landscapes to meet unfettered growth in demand for natural resources, which is inconsistent with movements towards sustainability, resilience and smarter use of resources and technology that current trends suggest will predominate in London in 2062.

## Surface water

Changing patterns of precipitation and population over the next fifty years in London will also affect surface water runoff and flooding. More intense rainfall on paved surfaces without remedial measures will lead to more frequent localised flooding as water is unable to drain away or local drains overflow. In central London the combined sewers constructed in the second half of the nineteenth century will overflow into the Thames Tideway Tunnel, which should have been operational for more than 40 years by 2062.

The Thames Tideway Tunnel will be one of the most significant infrastructure projects in London in the next decade. It is being constructed to stop sewage overflowing into the Thames during high rainfall events. The Tideway Tunnel was chosen in the 2000s as the most cost effective solution to an environmental problem caused increased runoff and the inherent design of London's Victorian sewerage system. In the 1850s the most cost effective option for solving the environmental and sanitary crisis facing London was to connect households to existing surface water drains, and to intercept both rainfall and sewage, which was then flowing directly into the Thames, in large intercepting sewers that run west-east across the city (Halliday, 2001). The system design, led by Sir Joseph Bazalgette, allowed for the intercepting sewers to overflow into the Thames in times of high rainfall, to prevent sewer flooding in homes and streets. Over time the frequency of these 'combined sewer overflows' (CSOs) has increased, due to the increasing impermeability of London's surfaces, greater flows of sewage from a higher population and potentially higher frequency of intense storm events due to climate change. Overflows now occur on average fifty times a year, polluting the tidal Thames.

The Tideway Tunnel will intercept all CSOs currently discharging into the Thames. The tunnel will follow the route of the Thames from west London to Tower Hamlets, then on to Abbey Mills pumping station near Stratford, and finally to Beckton Wastewater Treatment Plant. It will be 32km long, approximately 7m in diameter, and up to 66m below ground. The tunnel will store dilute sewage that currently overflows into the Thames, and it will be pumped out and treated at Beckton. The Tideway Tunnel is effectively an extension of Victorian infrastructure design, being the ultimate interceptor for Bazalgette's intercepting sewer system.

Alternatives to the Tideway Tunnel, such separating surface and wastewater, or increasing the permeability and local storage of surface water across London, before it enters the sewers, were dismissed by Thames Water as being too costly and unreliable. They also represent a radical departure from how surface water is managed in London. Building concrete tunnels and pumping stations, and expanding sewage treatment works are consistent with how surface water and sewage have been managed in this city since the Great Stink of 1858. In building the Tideway Tunnel 'super sewer' the water industry is doing what it knows best. However, we may have missed an opportunity to transform London's environment, buildings, streets and waterways.

The key alternative to the Tideway Tunnel promoted by its opponents is based on the principles of Sustainable Urban Drainage Systems (SUDS). SUDS aim to manage surface water where it falls, and to store water locally, rather than discharging immediately to sewers or the environment. Such measures include green roofs to absorb rainfall, rainwater tanks to store water, ponds and swales to store water in green spaces, permeable paving, and any measures that encourage infiltration and increase the capacity for the urban environment to store and treat surface water (Woods et al, 2007). These measures often deliver multiple benefits – rainwater harvesting provides a source of water for non-potable uses such as toilet flushing; green roofs provide insulation benefits to buildings; ponds, swales and other measures also increase biodiversity, reduce the urban heat island effect and can enhance green spaces for leisure, learning and relaxation. Drainage and surface water management are a key benefit of what is known more generally as 'Green Infrastructure'.

'Green Infrastructure' was rejected in favour of 'Concrete Infrastructure' as the solution to the problem of combined sewer overflows in London in the 2000s. This was largely justified in terms of cost, but green infrastructure represents a more fundamental shift in how water is managed in London. Rather than surface water being understood as a public health and environmental hazard to be removed, contained and treated, green infrastructure solutions make space for water in the urban environment. They recognise the importance of water in restoring local ecosystems and habitat, providing green space for people and relieving the urban heat island effect. However, they are more difficult to control than concrete infrastructure solutions. One big sewer is much easier for a large utility to manage than thousands of small interventions in private and public spaces and buildings across the city. The Tideway Tunnel solution affirms nineteenth century models of managing and governing water infrastructure, as much as it is a continuation of Bazalgette's interceptor design principles.

The operational success of the Tideway Tunnel will be dependent on the cost of energy. The energy requirements for pumping and treating the additional volume of sewerage will be significant, and may become prohibitively expensive if energy prices increase dramatically. It is possible that by 2062 the Tideway Tunnel will be too costly to operate, and 'low energy' surface water management options, such as SUDS, will be ever more important. It may be energy shortage, rather than environmental or public health crises, that finally dislodges nineteenth century approaches to infrastructure in London.

It is therefore important for engineers, planners and urban designers in London to continue to pursue green infrastructure solutions to surface water management over the next fifty years, despite the presence of the Tideway Tunnel. Green Infrastructure delivers solutions to multiple environmental and social problems, as well as low energy surface water management. Although not convinced of their ability to address the scale of the problem of CSOs as reliably and cost effectively as the Tideway Tunnel, Thames Water have been supportive of SUDS in general. However, Thames Water remains fundamentally a 'Concrete Infrastructure' company, who primarily exist to provide clean water and remove dirty water in London.

Green Infrastructure delivery will ultimately be the responsibility of a wide range of actors, including local authorities, the Greater London Authority, the Environment Agency, central government, developers, housing providers, building owners, community groups, non-governmental organisations and citizens. Green Infrastructure distributes water and habitat across the city, and it distributes ownership and responsibility. Developing new governance and ownership models for Green Infrastructure will be essential in achieving the multiple benefits it can deliver in solving key environmental and resource challenges. If Green Infrastructure approaches are successful, London in 2062 may be a cooler, greener, more pleasant city, with healthy local waterways and thriving wildlife, than will be the case if Concrete Infrastructure continues to prevail.

## Flooding

The greatest risk of flooding in London comes from the sea. London is situated on the Thames Estuary and at risk of flooding from storm surges in the North Sea, which can be particularly devastating if they coincide with spring high tides, as was the case in 1953 when 307 died in a flood in the Thames Estuary and other parts of southern England (Thames Estuary Partnership, n.d). Currently the land at risk of flooding in the Thames Estuary contains 1.25 million residents, eight power stations, 35 tube stations, 167km of railway line and property valued at £200 billion (Environment Agency, 2012).

Constructed flood defences have a long history in London, but the most iconic feature of the city's flood defence system is the Thames Barrier, which has been operational since 1982. The construction of the Thames Barrier was the major outcome of a review of flood defences following the 1953 flood, and its future operation is one of the key issues to be addressed by 2062.

The sea level in the Thames Estuary is currently rising at 3mm per year, a result of climate change and sinking land (Environment Agency, 2012). The south of England has been sinking and the north of Scotland rising since the end of the last ice age, rebounding after the loss of the weight of ice on the land in the north. This accounts for about 1.5mm rise in relative sea level, with the remainder a result of thermal expansion of the oceans and glacial melting due to climate change (Environment Agency, 2012). By 2060 sea levels in London are forecast to be 22.2cm higher than 1990 levels under a low emissions climate change scenario, and 31.4cm higher under a high emissions scenario (Millin, 2010). The worst case, and highly unlikely scenario for London sea level rise, considering high levels of uncertainty in the contribution of polar and glacial ice melting on global sea levels, is 2.7m by the end of the twenty-first century (Environment Agency, 2012).

Between 2002 and 2012 a major review of flood management in the Thames Estuary was conducted by the Thames Estuary 2100 (TE2100) project. This review was centred on climate change forecasts for flood risk in the estuary, but also considered economic, social and demographic change, and the need to maintain habitat for wildlife. The project reviewed the performance of existing flood defences in the light of climate change forecasts and developed an action plan to address current and future risks. The plan was designed to be technically feasible, adaptable to change, environmentally sustainable, economically justifiable and socially and politically acceptable.

The key recommendations of TE2100 are that the current system of flood defences should provide adequate protection from flood risk until 2070, requiring on-going investment in maintenance until 2035, and major investment in replacement and upgrade after that (Environment Agency, 2012). The relatively long life of this protection is the result of lower rates of sea level rise than the 8mm per year that the defences were designed for. The TE2100 plan makes some assessment of options for major new investments to provide protection after 2070, the most promising of which is a new barrier across the Thames at Long Reach. By 2062 planning and construction of this new barrier, or an alternative option, should be well underway, as the current defences will be nearing the end of their effective life. The plan also addresses the need to create new habitat for wildlife in the Thames Estuary, as existing habitat sites between flood defences and the low tide water level will be squeezed out by rising sea levels.

Climate change is also likely to result in an increase of fluvial (from rivers) and pluvial (from surface water) flooding in London. Increased winter rainfall could result in up to 40% increase in peak fresh water flows at Kingston by 2080 (Environment Agency, 2012). The Thames Barrier, designed to protect against tidal surges, has been used increasingly to managing fluvial flood risk in recent years, and together with flood walls and other defences is likely to be used more frequently for this purpose. Managing pluvial flood risk is an important function for urban drainage, and over the next fifty years should be integrated with green infrastructure and SUDS implementation.

Flood defences are only one element of flood risk management. Preparing for flood events through good design and emergency planning will also be of increasing importance in 2062. Land use planning to avoid high risk development in flood plains is also vital. Vulnerability to flooding is a combination of land use planning, preparedness, and hydrology. Critical infrastructure, hospitals, aged care facilities, schools and police stations should not be planned in high flood risk areas, and by 2062 it may be possible to relocate existing facilities away from these areas. Building design in flood risk areas over the next fifty years will include resilience and local defences, such as household scale flood barriers and waterproof or low value uses on ground floors.

The TE2100 plan is notable in calling for integration between government agencies and local government authorities, as well as participation from key stakeholders including infrastructure providers, key business interests, NGOs and local communities. The plan is designed to be adaptable, subject to review every ten years and major revision in 2050 to develop more detailed plans for the end of the century. The plan outlines ten key indicators to be monitored to evaluate the on-going effectiveness of existing defences and to reduce uncertainty of key parameters such as sea level. The success or failure of this approach to planning will be evident in 2062, when the detailed planning and implementation of new defences for conditions beyond 2070 will need to be in place.

## Conclusion

London has been well served by water and drainage infrastructure built in the nineteenth century and flood defences constructed in the twentieth century. These systems leave technical, environmental, economic and institutional legacies. The extent to which London is 'locked-in' to particular modes of infrastructure provision depends on a range of social and political factors, as well as the technical possibilities left open by the pipes, drains and flood walls that have been built over the last 200 years. The next fifty years will present opportunities to develop new models of the provision of water infrastructure services that take advantage of new technologies and governance structures, and reflect new understandings of the relationship between the city and the landscapes that sustain it.

# References

Bell S. and Aitken, V. 2008. The socio-technology of indirect potable water reuse. Water Science & Technology: *Water Supply.* 8: 441-448.

Bell S., Shouler M., Tahir S. and Campos L. 2013. Integrating social and technical factors in decision support for non-potable water reuse networks. Paper presented at Asia Pacific Water Recycling Conference. 2 July. Brisbane

Butler D. and Memon F. 2006. *Water demand management.* London: IWA Publishing

DCLG. 2010. *Code for Sustainable Homes Technical Guide.* London: Communities and Local Government Publications. Available from: https://www.gov.uk/government/publications/code-for-sustainable-homes-technical-guidance. [Accessed 21 June 2013]

DEFRA. 2011. *Water for Life: the Water White Paper.* London: The Stationery Office

Environment Agency. 2008. *Water resources in England and Wales – current state and future pressures.* Bristol: Environment Agency. Available from: http://cdn.environment-agency.gov.uk/geho1208bpas-e-e.pdf. [Accessed 21 June 2013]

Environment Agency. 2012. *TE2100 Plan Thames Estuary 2100.* London: Environment Agency. Available from: http://www.environment-agency.gov.uk/homeandleisure/floods/125045.aspx. [Accessed 21 June 2013]

Greater London Authority. 2012. Population Projections to 2041 for London Boroughs by single year of age and gender using the Strategic Housing and Land Availability Assessment (SHLAA) housing data. Available from: http://data.london.gov.uk/datastore/package/gla-population-projections-2012-round-shlaa-borough-sya. [Accessed 16 August 2013]

Hartley T. 2006. *Public perception and participation in water reuse.* Alexandria, USA: The Water and Environment Research Foundation.

Halliday S. 2001. *The Great Stink of London.* Abingdon: The History Press

Knight H., Maybank R., Hannan P., King D. and Rigley R. 2012. The Old Ford Water Recycling Plant and non-potable water distribution network. *Learning Legacy.* London: Olympic Development Authority. Available from: http://learninglegacy.independent.gov.uk/documents/pdfs/sustainability/old-ford-case-study.pdf. [Accessed 21 June 2103]

Millin S. 2010. UKCP09 sea level estimates. UKCIP Briefing Notes. Available from: http://www.ukcip.org.uk/wordpress/wp-content/PDFs/UKCIP_sea-level.pdf. [Accessed 21 June 2013]

Nickson A., Tucker A., Liszka C., Gorzelany D., Hutchinson D., Dedring I., Reid K., Ranger K., Clancy L., Greaves M. and Wyse N. 2011. *Securing London's Water Future.* London: Greater London Authority. Available from: http://www.london.gov.uk/sites/default/files/water-strategy-oct11.pdf. [Accessed 21 June 2013]

Parkes C., Kershaw H., Hart J., Sibille R. and Grant Z. 2010. *Energy and Carbon Implications of Rainwater Harvesting and Greywater Recycling. Final Report.* Bristol: Environment Agency

Thames Water. n.d. Our draft Water Resources Management Plan 2015-2040. Available from: http://www.thameswater.co.uk/about-us/5392.htm. [Accessed 21 June 2013]

Thames Estuary Partnership. n.d The Thames Estuary Floods 1953. Available from: http://www.thamesweb.com/1953-floods.html. [Accessed on 21 June 2013]

UKCP09. n.d. UK Climate Projections. Available from: http://ukclimateprojections.defra.gov.uk/. [Accessed 17 August 2013]

Waterwise. n.d. At home. Available from: http://www.waterwise.org.uk/pages/at-home.html. [Accessed 21 June 2013]

Woods Ballard B., Kellagher R., Martin P., Jefferies C., Bray R. and Shaffer P. 2007. *The SUDS Manual.* London: CIRIA

# Hydro-urban London

## Tse-Hui Teh

**Figure 1:** Photomontage of new infrastructure spaces for water reuse, energy production, food production, biodiversity and recreation.

**How to cite this book chapter:**
Teh, T. 2013. Hydro-urban London. In: Bell, S and Paskins, J. (eds.) *Imagining the Future City: London 2062*. Pp. 95-96. London: Ubiquity Press. DOI: http://dx.doi.org/10.5334/bag.m

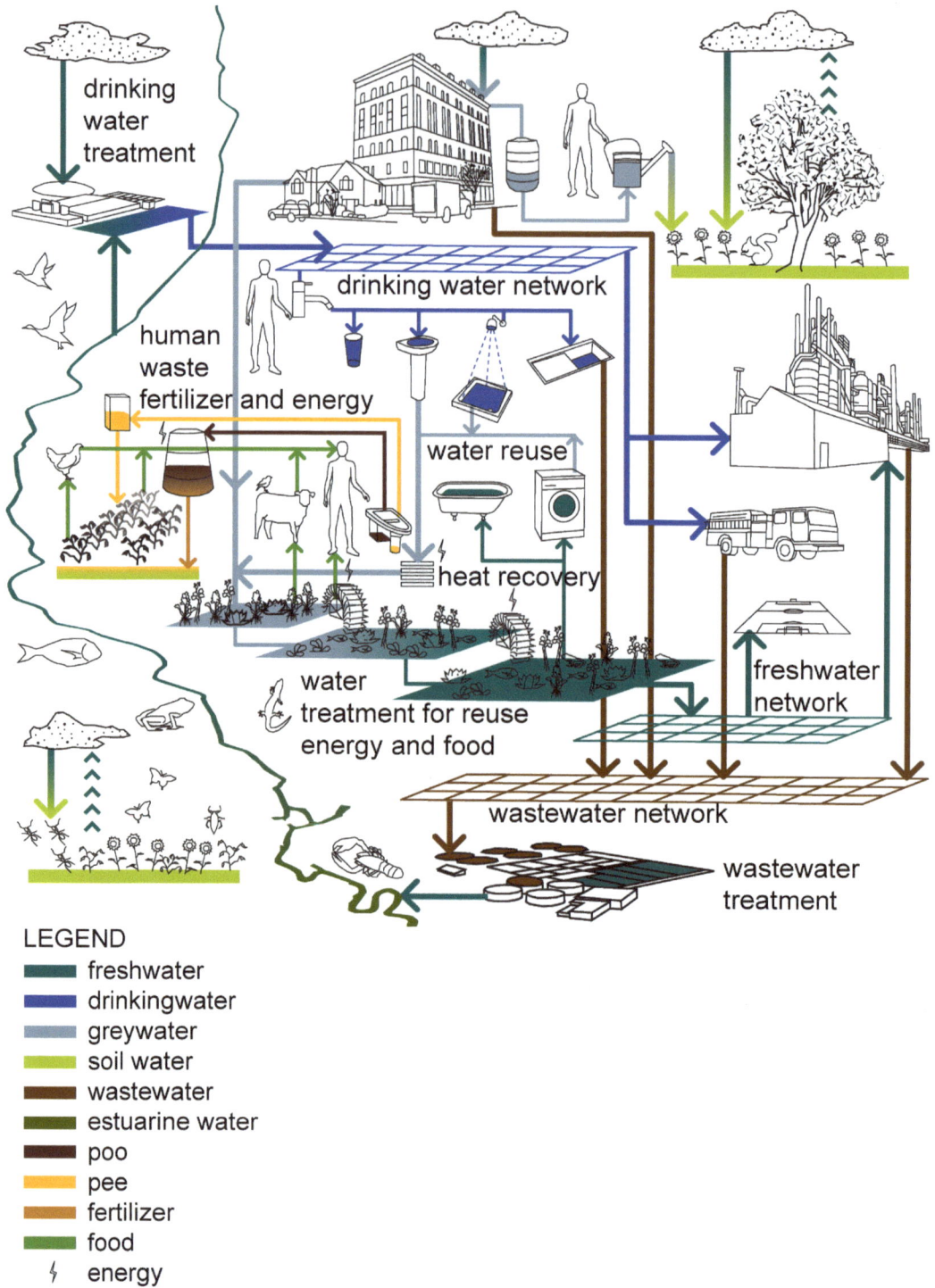

**Figure 2:** Diagram of water, food and fertilizer flows for the new infrastructure spaces of water reuse, energy production, food production, biodiversity and recreation.

# The future of food

## Robert Biel

## Introduction

How to feed the city? This has been a predominant question during recent centuries of human development. Conventional solutions are revealed as unsustainable in the long term: massive chemical inputs, factory farming and agribusiness, global food imports. A new approach must therefore be found. This chapter addresses the scope of producing food *in* London itself. As a conceptual approach, we define a 'backcasting' methodology and outline the different levels and relationships of an urban food system, as well as the wider parameters which form its context: the revolution of rural agriculture, and the solar transition. We propose a threefold categorisation, namely the urban forest, subsistence plots, and ultra-high productivity farming. In each case, we illustrate the potential from existing experiences, while being aware that the emergent possibilities of the new ensemble will far outstrip the sum of their parts.

## Framework and methodology

We could interpret 'food in London' to mean a situation where London, as object, gets food (sustainably) delivered *to* it. Here, I take an opposite approach: London's people as *subject*, possessing food autonomy, control over land and knowledge. Currently, this concept would be named 'food sovereignty' (Pimbert, 2009), although there are on-going debates which may re-define or replace this term eventually.[1] Although food *consumption* (and the interactions arising through it) would be an important topic, I will focus on production.

My methodology is similar to backcasting: visioning a desired outcome and building back from it. This differs from forecasting or scenario-building, both of which build forward from the present (Wearerising, n.d). We don't therefore detail the *un*desired (e.g. Mad Max-style) futures

---

[1] Developments in Mexico suggest, for example, that the term may be subject to co-optation by ruling interests: Rita Pérez, personal communication.

**How to cite this book chapter:**
Biel, R. 2013. The future of food. In: Bell, S and Paskins, J. (eds.) *Imagining the Future City: London 2062.* Pp. 97-106. London: Ubiquity Press. DOI: http://dx.doi.org/10.5334/bag.n

which could follow ecological crisis. Nevertheless, the tasks of *avoiding* these will be addressed. Though this vision is personal, it needn't be merely subjective: we will be identifying trends within today's food-related experiments as future growth-points; while insights from general systems theory can help us understand how systems evolve and acquire resilience.

We can't analyse food separately from wider changes; here, we assume three:

1. Today's political, economic, social and institutional crisis has been resolved in favour of radically increased equity and community control.
2. Society adapts to life without fossil fuels.
3. Climate adaptation and mitigation are central concerns.

While these three conditions are *broader* than the food issue, they're not *external* to it:

1. The bottom line for socio-politico-economic futures is to feed the people, and it is precisely over land and food that many key struggles for rights are focused.
2. Food growing is the most basic way to harness solar energy, and a key area for minimising emissions.
3. It is inseparable from both the problems of and responses to climate change (combatting erosion, conserving water).

The technical shift to solar economy, and societal restructuring, will occur through mutual interaction (Schwartzman, 2009). London's climate will be different in 2062, most likely warmer, though possibly colder; but at least we're certain there will be less stability and more extreme events. Within a given year,[2] it will be warm or cold at the 'wrong' times, confusing plants and animals. So adaptability and resilience are key.

## A systems approach to food

In visualising a systems approach to food, it is helpful to think of an 'urban metabolism', whereby flows – of resources, of 'waste' from one process becoming an input for another, of water – serve as a physical medium for networks and transactions linking people and communities. We must also address the time dimension, through which systems evolve. In this respect, systems adapt to step level changes, which on the one hand confront them as challenges (extreme environmental events, for instance), but which on the other hand they learn to *embrace*, thus making radical change an integral mechanism of their own development: this is the meaning of 'transition'.

Food systems occur at different 'levels' or scales, nested within each other; networks or 'chains' run through and between them. London lies within national and international food chains, and is internally differentiated into Greater London, Inner London, and the community/locality. Our system rule is that exchange is good, provided it minimises entropy, i.e. exports of disorder. No level should build its order at the expense of degrading either its surrounding system ('environment', physical or social), or any excluded and super-exploited regions.

What we must escape *from* is today's situation, where food security in the global North is achieved by degrading the environment of the South; everything is achieved by degrading the soil; urban food systems squeeze the farmer through corporate dominated chains; and poorer communities suffer deprivation. There is an intrinsic link between injustice and wastefulness: the process of exploitation is simultaneously a loss of quality (nutritional quality [Caldwell, 1977]), but also a negative Energy Return on Energy Invested (Glaeser & Phillips-Howard, 1987). Rebellions

---

[2] This was already the case in 2008 and 2011.

of the oppressed thus reduce entropy. The repercussions of such events will be worldwide. We assume that today's exploitative world food system – where extreme concentration of knowledge and landholding obliges farmers to export food at the expense of local staples – will have been overthrown by grassroots food-sovereignty struggles in the global South. London's local food system will have changed step by step against this background. The apparent 'constraint' of not being *able* to import food will, in a positive sense, bring about a transformative rise of creativity.

The resultant systems will remain interlinked through (non-exploitative) chains and networks, which stimulate emergent properties of the whole. At the same time, there's a strong dose of local self-sufficiency (within each community, within Inner London, within London relative to wider systems). This is *partly* to minimise food miles, but mostly to gain resilience to shocks: where higher-level structures are trashed by extreme events, systems must regenerate from anywhere (Brafman & Beckstrom, 2006).

Growing technique and institutions are analogous: each minimises energy/input. So, just as low-input farming accesses the soil's self-organising properties, institutions draw free energy from human resourcefulness. Imagining networks as *rhizomes* (Deleuze & Guattari, 1987) further emphasises the analogies. Much of systems thinking on food (call it agroecology, permaculture or whatever) envisages changing farming and society in tandem; in both cases, creativity springs from shunning too much equilibrium, focussing on the *edges*, the margins (Chinmay, 2009; Whitefield, 2004).

Comparing different levels, we find certain 'isomorphisms' (structural similarities) between them. Thus, institutionally, commons regimes work equally well whether at intimate micro-levels (for a tangible resource like a piece of land); or globally, as in the infosphere, through open source knowledge. With increased co-operation a logical response to crisis (Nowak, 2006), regimes undergo resurgences, forming the major single principle of organisation at each level.

Do we find such isomorphisms in the food growing sphere too? Only partly. All levels are similar in pursuing sustainable principles:

1. conserving soil structure
2. diversity of crops, and of strains within each crop
3. biodiversity in a larger sense, i.e. working with wildlife, pollinating insects, natural predators
4. working with plants' natural resilience.

These embed complex order (resist entropy) in the land itself. But food systems may be radically *dissimilar* as we change levels, especially between the urban and the rural; with peri-urban farming sharing features of both.

## Components of urban agriculture within a wider food ecology

In a static sense, Britain could feed itself *now* (Fairlee, 2007). But in our longer fifty-year time-scale, we must halt long-term degradation trends. For this, drastic changes are needed. Some are organisational: plantation/agribusiness gives way to something else, perhaps smaller farms (Rosset, 1999), accompanied by de-urbanisation. But more profoundly, humanity must, by 2062, have resolved today's threatened disappearance *of the soil itself* (Hough, 2010), an issue sometimes called 'peak soil' (Montgomery, 2008). Soil now vanishes at up to fifty tonnes per hectare per year, 100 times faster than it is formed (Banwart, 2011). With soil conservation 'central to the longevity of any civilization,' (Montgomery, 2007) this already impacts on UK policy (BBC, 2009). The answer lies in linking carbon sequestration and soil fertility: a benign positive feedback loop, since high carbon-content soil promotes more growth and thus more sequestration (Brown, n.d). Soil holds nearly three times as much carbon as vegetation and twice that of the atmosphere, the use of no-till agriculture maximising its potential in this respect (Wang et al, 2011).

Climate mitigation and feeding the people should therefore be 'win-win', but how to kick-start it? Today's ideas include large-scale ranges where animal grazing acts as a carbon pump (Norman, 2001; Savory, 1983); or a charcoal based method replicating the 'terra preta de indio' of the ancient Americas (Taylor, 2010). These are big debates, outside our scope here. Suffice it to say, there will be huge changes in rural farming, within which is inserted an urban system, probably very different from the rural one, but equally radical in the changes it implies. The latter is our focus here.

Such changes, meeting the goals of equity, dis-alienation from nature and energy descent (a term used in the Transition movement to denote the steps leading to a 'visioned' low-energy future), liberate the city's genius for self-organisation. Farming and built environment are no longer sharply separate (Wilson, 2009). Food growing takes many forms. Though these shade off into one another, to simplify, I'll propose three categories:

a) the urban forest
b) subsistence plots
c) ultra-high productivity farming.

Firstly, in relation to what I am calling the urban forest, 'greening' the city will be thought of differently, mainly in food terms. Transition models rightly insist that we consider trees as producers of fruit and nuts (Hopkins, 2008), a strategy already underway in London (The London Orchard Project, n.d). Not only will edible urban forests, once established, have their own self-maintaining ecology (Ettinger, 2012), but the process of creating them is *itself* emergent, a spontaneous encroachment of growing spaces, as already foreshadowed by the squatted community of Bonnington Square, Vauxhall (Self-help-housing, n.d). Guerrilla gardening (Reynolds, 2008), referencing guerrilla as a diffuse, self-organising form, is a societal struggle conducted through the self-organising capacity of nature: as in its adaptation of Masanobu Fukuoka's seed-balls (whereby plants themselves choose where to grow) as 'seed-bombs'.[3] Hidden rivers like South London's Effra will be re-created, with re-established habitats redressing today's deficit of natural predators like frogs and hedgehogs. Through the networking of green fingers, agriculture becomes seamlessly part of the city.

Secondly, let us consider how urban subsistence farming may impact food security. Between the wars, perhaps 700,000 tonnes of vegetables grew in English and Welsh allotments, numbering about a million plots[4] (Acton, 2011). This might feed four million people. The volume is encouraging. But top-down strategies, placing food security within (military) national security, pushed allotments only in short bursts, typically wartime, gaining a strong initial output through chemicals only for diminishing returns to set in.[5] We, on the contrary, aim for high production in the long term, 2062 *and beyond*. For this, the soil's own structure and fertility must always be renewed.

This we achieve through organic mulches[6] (Dowding, 2007). Taking the traditional allotment (250m²), converted to a no-dig method with paths between beds, our cultivable surface is about

---

[3] As a variant of this, 'Guerrila Grafters' in San Francisco are grafting fruit-bearing branches onto ornamental cherry, plum and pear trees (Zimet, 2012).

[4] Figures are from 1923, and include 200,000 tonnes of potatoes.

[5] In 1944, there was an attempt to convert the wartime 'Dig for Victory' slogan into a more permanent 'Dig for Plenty' (http://www.nationalarchives.gov.uk/theartofwar/prop/production_salvage/INF3_0098.htm); however it lacked a clear orientation and soon fizzled out.

[6] The mulch also shields the soil from erosion, prevents water loss, and suppresses weeds.

150m². Allowing for a typical 40mm mulch (Corbalan, 2005), over 150m² this gives 6m³. But, only half is internally generated by the plot.[7] This gives a figure for what we require from the urban metabolism, in the form of compostable waste.[8] Where today's visioning of metabolism, as industrial ecology or industrial symbiosis,[9] tends to view agriculture peripherally, as an outlet or sink; in that of the future it will be central.

Significant scope exists to expand urban subsistence farming. In today's Elephant and Castle, potentially cultivable land could maybe yield 26% of the population's vegetable needs[10] (Tomkins, 2009). But this doesn't necessarily mean more conventional allotments. While I hope there'll still be a role for them (allotments *have* kept alive commons regimes against all odds!), the new urban farmland will be institutionally innovative: community land trusts (Davis, 2010) could be a starting point.

Thirdly, to complement the previous two categories, we assume an ultra-high productivity sector, but what form will it take? We must unquestionably embrace cutting-edge science in this respect. A key issue remains, however: how far can or *should* we try to free ourselves from nature? There's a historical legacy of false aspiration to control/circumvent nature (Merchant, 1990). The above question could further be resolved into two aspects, [i] *energy medium* and [ii] *growing medium*.

Lighting by light-emitting diode (LED) already makes possible highly energy efficient growlights in a mixture of red and blue wavelengths (used experimentally with success by the author); given LED's current exponential trajectory of efficiency gains, we can confidently predict by 2062 a technical revolution opening up the possibility of growing food inside buildings (either multiuse ones, or ones built specially). Futuristic visions often assume this (Despommier, 2010). As we write, these move a step closer, with 'plantscraper' models on the point of realisation (Ma, 2012). Nevertheless, there are problems: most obviously, since the only fundamental defence against entropy is drawing energy from the sun (Georgescu-Roegen, 1975), the most straightforward way is still to expose plants to it directly. More subtly, we respect the principle of working with/like nature, one aspect of which is plants' natural defences: their biorhythms being synchronised with that of insects (Solon, 2012); artificial light could pose problems. This might be addressed through the solar energy medium combined with non-soil growing medium, as in hydroponics... but not fully. Plants still need soil in order to 'network': for example, only by chatting through fungal and mycorrhizal filaments can they fully trigger pre-emptive responses to disease (Song et al, 2010). Such information-exchange should not surprise us. In organic farming theory, humus is a complex system in its own right (Howard, 1943), and in complexity theory, we re-phrase entropy in information terms (Morin, 2008). Thus, to safeguard soil against degradation (structure-loss) means also to nurture its capacity for information storage and transmission.

This makes me doubtful about going too far in 'escaping' nature. But, given London's latitude and food crops' need for light, a more conventional, *but intensive*, soil based, natural light system can flourish if we move upward. The key area is rooftops, with greenhouses insulated, not heated. The SolaRoof model is interesting here: because open source (SolaRoof, n.d), it anticipates a wider flourishing of knowledge commons.

But the point is not to *restrict* technical innovation in the ultra-productive sector. On the condition that its own self-organising principles interlock with those of nature, it can be as daring as it pleases, through hydroponics or whatever. Today we see this in the work of Will Allen in Milwaukee (Herzog, 2010), where a multi-layer aquaponic greenhouse system, heated by com-

---

[7] Author's research.

[8] A similar calculation could be made for water.

[9] See www.tees.ac.uk/docs/DocRepo/Clemance/IndustrialSymbiosis.pdf for a typical diagrammatic presentation.

[10] Calculated on the basis of a population of 16,245, using UK weekly vegetable consumption of 1,600 g, giving an estimated potential yield of 267 tonnes.

posting, carries water populated by fish and purified by the plants themselves: a spontaneously self-regulating system.[11]

In relations between the different sectors and levels, there will be some specialisation. Limits to this are posed by the resilience condition (with extreme events, climate fluctuation and disturbed seasonality the norm, each plot must have some self-sufficiency, spreading its options by growing many crops and strains). Nevertheless, specialisation between peri-urban and central London are key; this dialogues with the garden city and green belt traditions in British planning history. Along lines pioneered by Community Supported Agriculture, and specifically London's Hackney Growing Communities (Growing Communities, n.d), peri-urban farms grow staples like cereals, while other crops (salads, or those like peas and maize, where sugars rapidly convert to starch) grow closer to the point of consumption.

The above are some of the elements. Where it gets exciting is the emergent properties of the complex whole. The mechanism for such emergence is networking, a rebuilding of society's information content. We see this just beginning now with the rise of new social and productive relations, at the same time highly concrete and linked intangibly through the infosphere (Local Harvest, n.d). Such spaces, or "islands of unpredictability" (Carlsson, 2008), will by 2062 have coalesced into emergent forms which can barely be envisaged today.

## Pathways of transition

The identifying feature of backcasting is starting from the future. Clearly, however, where we'll be in fifty years depends on how we get there: that's the issue of transition. Today's transition models arguably underestimate the radicality of *social* changes needed (Biel, 2010). It's true we're *defending* structure, both of soil and of society, against the entropy threatening to engulf it, and therefore logically pattern social organisation on nature. Nevertheless it's too easy to dress up existing hierarchies under a semblance of biomimicry. Futures thinking should, in pursuit of new paths, accept the need to 'break' existing path-dependencies (Tiberius, 2012), which implies identifying correctly what these are. What's causing today's food crisis is inseparable from a certain dangerous momentum from within the mode of production itself, which currently degenerates in two complementary/contradictory ways, both of them depleting order: [a] a concentration of power at the top; [b] a chaotic disaggregation, parasitised upon by finance capital (Biel, 2012). Both are strongly present in the food system: dis-embedding knowledge, appropriating genetic resources, land grabs (driven equally by militaristic food security projects and chaotic speculation). If we're to have any future, humanity must conquer this degeneration, a process which must build upon the forces *now* resisting it.

To understand these forces is to understand future institutional structures. The roots lie in the global South, most easily traceable to the late 1990s in the impact of the Mexican Zapatistas and Indian farmers' movements (Ainger, 2003). This tradition of struggle tends to surface in a succession of forms, which have recently included a Land and Freedom camp (October 2011) on London's Clapham Common (which referenced global struggles (Heggs, 2011), as well as the Diggers' 1649 occupation in what is now London's outskirts) and shortly afterwards the Occupy movement, critically highlighting issues of land/space and which included Occupy Our Food Supply (Occupy Our Food Supply, 2012). However ephemeral its specific forms, we can reasonably see this as a coherent current of struggle likely to grow in strength.

The right to land has never been wholly separable from how that land is managed: distribution in isolated plots provides little defence against predation, whereas commons might. Appropriately, 2012 is the year of co-operatives (United Nations, n.d); they have momentum on their side,

---

[11] Author's notes on a presentation by Will Allen, London 2008.

with co-ops now quietly challenging the conventional economy (Bollier, n.d). In today's London, OrganicLea provides maybe the best example of what could be achieved (Organic Lea, 2013): the conditions for *replicating* this are already under debate (Reclaim the Fields, n.d). The struggles in which they are born thus influence the institutional forms of London's food-growing future, the restructuring of society itself co-evolving with the systems by which it is nourished.

## Conclusion

The future of food in London will be one where challenges will have been embraced as opportunities, in which the self-organising properties of society complement and intertwine with those of nature. The city will in one sense appear 'wilder', with a new generation of green buildings and green corridors in which self-balancing populations of birds, insects and small vertebrates control what were once considered 'pests'. At the same time, a high-efficiency urban farming will employ biomimicry approaches like intercropping to operate through patterns of interdependent diversity, replicating those of natural systems. Scientific research will both facilitate and draw data from the constant experimentation of practical food-growers, while in an institutional sense, new property relations in land and space, such as commons regimes and community land trusts, will open up society's creative potential. A snapshot of 2062 will therefore above all reveal a city in the midst of a many-sided adaptive and proactive process of emergence and development.

## References

Acton L. 2011. The Allotment Movement in North-East Greater London 1900-2010. Thesis (Ph.D.), University College London

Ainger K. 2003. Life is Not Business: the intercontinental caravan. In Ainger K., Chesters G., Credland T., Jordan J., Stern A. and Whitney J. editors. *We Are Everywhere*. London: Verso; p. 160-170.

Banwart S. 2011. Save our soils. *Nature*. 474: 151-152.

BBC. 2009 (24 September). Bid to Protect England's Topsoil. Available from: http://news.bbc.co.uk/1/hi/sci/tech/8272022.stm. [Accessed 12 August 2013]

Biel R.A. 2010. Towards Transition. In: Levene M., Johnson R. and Roberts P. editors. *History at the end of the world? history, climate change and the possibility of closure*. Humanities Ebooks

Biel R.A. 2012. *The Entropy of Capitalism*. Leiden: Brill

Bollier D. n.d Biggest Secret in the World Economy - Cooperatives employ more people than multinational companies. Commons Magazine. Available from: http://www.onthecommons.org/biggest-secret-world-economy. [Accessed 12 August 2013]

Brafman O. and Beckstrom R.A. 2006. *The Starfish and the Spider – the Unstoppable Power of Leaderless Organizations*. New York: Penguin

Brown S. Connections between Compost, Soil Carbon and Climate Change. Available from: http://www.agresourceinc.com/article_connections_print.htm. [Accessed 12 August 2013]

Caldwell M. 1977. *The Wealth of Some Nations*. London: Zed Books

Carlsson C. 2008. *Nowtopia*. Oakland: AK Press

Chinmay. 2009 (17 March). Permaculture Theme: Mind The Edge. Available from: http://sustainable-farming.blogspot.com/2009/03/permaculture-theme-mind-edge.html. [Accessed 12 August 2013]

Corbalan J. 2005. Compost de Broussailes. Abécédaire du Jardinier bio, *La Gazette des Jardins*, hors-série. October-December.

Davis J. 2010. *The Community Land Trust Reader*. Lincoln Institute of Land Policy

Deleuze G. and Guattari F. A. 1987. *A Thousand Plateaus: Capitalism and Schizophrenia.* Minneapolis: University of Minnesota Press

Despommier D. 2010. The Vertical Farm – Feeding the World in the 21st Century. Available from: http://www.verticalfarm.com. [Accessed 12 August 2013]

Dowding C. 2007. *Organic Gardening: The Natural No-Dig Way.* Totnes: Green Books

Ettinger J. 2012. Seattle Building Massive Edible Forest Filled with Free Food. Available from: http://www.organicauthority.com/blog/organic/seattle-building-massive-edible-forest-filled-with-free-food/. [Accessed 12 August 2013]

Fairlee S. 2008. Can Britain Feed Itself? *The Land.* 2007-8; Winter.

Georgescu-Roegen N. 1975. Energy and Economic Myths. *Southern Economic Journal.* 41(3).

Glaeser B. and Phillips-Howard K. 1987. Low-energy Farming Systems in Nigeria. In Glaeser B. editor. *The Green Revolution Revisited - Critique and Alternatives.* London; Allen and Unwin. p. 126-149

Growing Communities. n.d. Available from: http://www.growingcommunities.org/. [Accessed 12 Augsut 2013]

Heggs L. 2011. Land and Freedom Camp on Clapham Common – London. Available from: http://www.demotix.com/news/844031/land-and-freedom-camp-clapham-common-london. [Accessed 12 August 2013]

Herzog K. 2010 (29 April). Will Allen among 'Time 100: The World's Most Influential People'. Milwaukee Wisconsin Journal Sentinel. Available from: http://www.jsonline.com/blogs/lifestyle/92414184.html. [Accessed 12 August 2013]

Hopkins R. 2008. *The Transition Handbook – from oil dependency to local resilience.* Totnes: Green Books

Hough A. 2010. Britain facing food crisis as world's soil 'vanishes in 60 years'. *Daily Telegraph.* 3 February.

Howard A. 1943. *An Agricultural Testament.* New York and London: Oxford University Press

Local Harvest. n.d. Available from: www.localharvest.org. [Accessed 12 August 2013]

Occupy Our Food Supply. 2012. Letter of Support. Available from: http://ran.org/occupy-our-food-supply-letter-support. [Accessed 12 August 2013]

Organic Lea. 2013. Available from: http://www.organiclea.org.uk/. [Accessed 12 August 2013]

Ma J. 2012 (12 March). A 'Vertical Greenhouse' Could Make a Swedish City Self-Sufficient. Available from: http://www.good.is/post/a-vertical-greenhouse-could-make-a-swedish-city-self-sufficient/. [Accessed 12 August 2013]

Merchant C. 1990. *The Death of Nature.* New York: Harper

Montgomery DR. 2007. *Dirt – the Erosion of Civilisations.* Berkeley: University of California Press

Montgomery D. 2008. Peak Soil. *New Internationalist.* Issue 418.

Morin E. 2008. *On Complexity.* Creskill: Hampton Press

Norman S. 2001. New Grazing Theory Puts Roaming Cattle to Work. *LA Times.* November 18

Nowak M. 2006. Five Rules for the Evolution of Cooperation. *Science.* 8 December: 1560-1563.

Pimbert M. 2009. *Towards Food Sovereignty, Reclaiming autonomous food systems.* London: IIED

Reclaim the Fields, UK. n.d. Where Next For The Community Food Movement? Available from: http://www.reclaimthefields.org.uk/of-this-land-zine/where-next-for-the-community-food-movement/. [Accessed 12 August 2013]

Reynolds R. 2008. *On Guerrilla Gardening – A Handbook for Gardening Without Boundaries.* Bloomsbury: London

Rosset P. 1999. Small Is Bountiful. *The Ecologist.* 29(8)

Savory A. 1983. The Savory grazing method or holistic resource management. *Rangelands.* 5(4)

Schwartzman D. 2009. Ecosocialism or Ecocatastrophe? *Capitalism Nature Socialism.* 20(1): 6-33

Self-help-housing. n.d. Bonnington Square – London. Available from: http://self-help-housing.org/case-studies/bonnington-square-london/. [Accessed 12 August 2013]

SolaRoof. n.d. wiki homepage. [Accessed 12 August 2013]; Available from: http://solaroof.org/wiki/.

Solon O. 2012 (14 February). Plants Use Body Clocks to Prepare for Battle. Available from: http://www.wired.com/wiredscience/2012/02/plants-use-body-clocks-to-prepare-for-battle/. [Accessed 12 August 2013]

Song Y.Y., Zeng R.S., Xu J.F., Li J., Shen X., Yihdego W.G. 2010. Interplant Communication of Tomato Plants through Underground Common Mycorrhizal Networks. *PLoS ONE*. 5(10): e13324.

Taylor P. 2010. *The Biochar Revolution*. Global Publishing Group.

The London Orchard Project. n.d. Available from: http://thelondonorchardproject.org/. [Accessed 12 August 2013]

Tiberius V. 2011. Path Dependence, Path Breaking, and Path Creation: A Theoretical Scaffolding for Futures Studies? *Journal of Futures Studies*. 15(4): 1-8

Tomkins M. 2009 (June). The Elephant and the Castle: Towards a London Edible Landscape. *Urban Agriculture Magazine*. 22: 37-38. Available from: http://www.ruaf.org/sites/default/files/UAM22%20London%2037-38.pdf. [Accessed 12 August 2013]

United Nations. n.d. International Year of Cooperatives 2012. Available from: http://social.un.org/coopsyear/. [Accessed 12 August 2013]

Wang Y., Tu C., Cheng L., Li C., Gentry L.F., Hoyt, G.D., Zhang X. and Hu S. 2011. Long-term impact of farming practices on soil organic carbon and nitrogen pools and microbial biomass and activity. *Soil and Tillage Research*. 117: 8-16

Wearerising. n.d. Backcasting. Available from: http://wearerising.org/2009/01/13/backcasting/. [Accessed 12 August 2013]

Whitefield P. 2004. *The Earth Care Manual: A Permaculture Handbook*. East Meon, Hampshire: Permanent Publications

Wilson A. 2009. Growing Food Locally: Integrating Agriculture Into the Built Environment. *Environmental Building News*. 18(2). Available from: http://www.buildinggreen.com/auth/article.cfm/2009/1/29/Growing-Food-Locally-Integrating-Agriculture-Into-the-Built-Environment/. [Accessed 12 August 2013]

Zimet A. 2012 (4 October). Guerrilla Grafters: Undoing Civilization One Fruitless Branch At a Time. Available from: http://www.commondreams.org/further/2012/04/10-0. [Accessed 12 August 2013]

# Power

In *Power* we take a closer look at the institutions and the structures that influence London, and the relationship between political, commercial and civil society organisations and the levers of power.

The London Plan from the GLA is held up to scrutiny in terms of climate change, inequality and diversity. We are asked to consider what lessons can be learnt, including by those in power, from London's existing alternative economies, and the countries in the global south.

This section includes a description of the increasingly fractured relationships between governance, commerce and civil society in London. In the city described in *Power*, wealth has flowed only to a very few, increasing inequalities and concentrating power in the hands of business. This is a city which has risen to be an economic world leader, thriving against the backdrop of a neoliberal consensus that has pushed commercial interests to the fore.

As well as the future of the city's economy, this section also considers the future for London's housing. Finance, transport and housing are inextricably linked, and we are asked to contemplate the end of the property-owning democracy, and a future that sees hedge fund managers taking it in turns to make hotel beds. Our final chapter sees one man's home as his paranoid castle.

Our authors consider future implications for the city's power structures, describing possible futures for the institutions and individuals that make up the city. We are introduced to the ways in which we can subtly shift the balance of power, restructuring our behavioural norms, leading to a more sustainable city. This section sees arguments for a less masculine, more open city, whose citizens talk in terms of collaboration, rather than competition. The power of music is also considered: we are encouraged to sing, rather than shop, and shown a musical line that takes us from The Specials to Adele, via The Fleet Foxes.

*Power* introduces a range of possible futures where empowered, engaged citizens react to current austerity, and the challenges we can foresee in the future, with creativity and alternative thinking.

# Governing a future London: the city of any dreams?

Rob Pearce and Mike Raco

*To be a Londoner is always to take a chance: an instinct that manifests itself, materially, by an enormous addiction to lottery that makes Londoners the chief per capita financial gamblers in the world, and which proves itself, at more significant levels, by their perpetual willingness to take the most improbable human risks. And I believe the spirit of the ugly old indifferent capital encourages the presence of such people; and by its very incoherent informality, enables them to discover one another more freely and happily than elsewhere in our land.* (MacInnes, 1962)

As one of the first truly global cities, London has always been shaped by three factors: the structuring forces of the global, continental, and regional politics and economics; the resulting flows of people and ideas into and out of the city and the rapidly increasing degree of connectivity to other cities and related markets. In turn, these forces have been mitigated and managed through the ever-changing relationship between an array of institutions, including political, commercial, and civil society organisations. At different points in history the ability of those living in London to influence and control these institutions, and hence their own growth and prosperity and that of the city, has ebbed and flowed. These actors shaped, and continue to shape, London's political economy according to and in response to global economic, political and social forces. The production and reproduction of the city is a direct result of the production and distribution of goods and services it enables and manages. These are the key to the city's economic competitiveness and social progress. Therefore, the balance of forces between governance, commerce and civil society in the city, and their relative influence on its political economy, shaped its past, and will shape its future. Finally, it is the nature of these forces themselves that has tipped this balance in different directions over the decades. These forces are moulded by Londoners themselves who have always been 'willing to taking improbable human risks' to achieve their goals and who have always lived in a city that has a mysterious capacity to bring them together, 'to discover each other', in their struggle for the soul of the city.

**How to cite this book chapter:**
Pearce, R and Raco, M. 2013. Governing a future London: the city of any dreams? In: Bell, S and Paskins, J. (eds.) *Imagining the Future City: London 2062.* Pp. 109-117. London: Ubiquity Press. DOI: http://dx.doi.org/10.5334/bag.o

By the early 1960s the benefits of post war macro-economic policies had produced a city where living standards were relatively high and a greater degree of equality was being achieved than ever before. Economic progress was mirrored by progressive social policy, resulting in the 'swinging London' of the 1960s leading the world in popular culture. The 1970s and '80s brought a contraction in the city, with rising inequality and fractured relationships between governance, commerce and civil society institutions, as they were challenged by the global economic storms. The era of progressive consensus politics ended, to be replaced by a new set of norms founded on visions of a Global City. During the 1990s and first decade of the new century the city felt the full impact of neoliberal western capitalism. This restructured the balance of power within the city's institutional framework, most dramatically represented by the 'big bang' in the financial sector in 1988, placing global commercial interests in the driving seat, in the form of financial services weakening the city's governance and political structures and marginalising civil society.

The structuring forces of the global economy elevated London to the world's leading city. However, the shift in power to global commercial interests meant that the economic growth that followed flowed to very few, increasing inequalities and leaving the city even further divided than the '70s and '80s. The near collapse of the banking system in 2008 proved to be a tipping point in the relationship between governance, commerce and civil society, once again recasting their respective roles in remaking the city. It demonstrated that the current model of economic growth in the city is built on unsustainable foundations. It seems likely that despite the deluded optimism of hyper-globalist writers in the 1990s and early 2000s, an economy based on assumptions of constant expansion is likely to fail.

Between now and 2062, London, therefore, faces enormous governance challenges; and yet political control in the city has long been a source of ambiguity. On the one hand, London is the most governed city in the UK. It has a relatively strong Mayor, powerful state bodies, and a planning system that generates comprehensive spatial policy visions and plans. On the other hand, there are costs in being a capital city. Developments are often large-scale and driven by national, rather than local priorities. There are confusing divisions of responsibility between different tiers of government and the city has experienced a remarkable degree of corporate domination. This has been manifest in both the formal political sphere, in which the Corporation of London has become a vociferous voice for global business interests, and in the changing nature of the city's welfare services and infrastructure, large parts of which are now under the control of private companies, and outside of the direct control of government agencies. The political governance of the city, and the city's political class, has become increasingly distant from the communities it purports to serve. There are also questions over the extent to which London's social changes are sustainable in the long run. It is a city with long traditions of political diversity, radicalism and activism and the sheer plurality of its communities in social and economic terms, poses real challenges for future cohesion. Throughout its history it has been a focus for resistance movements, political protest, and violence. If current trends towards polarisation continue, it seems likely that there will be degree of 'backlash', in whatever form, with obvious implications for governance and democracy. The attributes MacInnes (1962) attaches to London provide some insight into this. In highlighting aspects prevalent in the 1960s that have become even more relevant to the contemporary city and its future, they provide an important additional perspective to the lens we will use to observe the city in 2062.

In this chapter we outline and discuss three alternative scenarios for the governance of the city over the next fifty years. We explore the first possibility that the problems highlighted by sustainability writers are overcome, and that the current model of globally-oriented growth continues. The result will be increased polarisation and rising inequalities. We will witness what Colin Crouch (Crouch, 2011) terms the growing corporatisation of public life in which the city's politics, welfare systems, and development trajectories become increasingly shaped by the activities of a post-political, post-democratic corporate elite. A second scenario is one of neo-Keynesianism, founded on the re-nationalisation and public acquisition of private interests in the name of a wider 'public interest'. We discuss the implications that this would have for the governance of the city and the likely position of elite groups of professionals and elected politicians. A third scenario is that a middle way emerges, in which a London of 2062 has been created via a new consensus capitalism – one which focuses on the relationships between societal and economic progress to create growth. We consider what would happen if capitalism were to be re-formatted using the concept of 'Shared Value', as developed by Michael Porter and Mark Kramer (Porter & Kramer, 2011). This might result in a reformed urban political economy, as a product of broader global struggles between political, commercial, and civil society institutions and actors. Global cities, such as London, will act as a battleground in which ideas over the future shape of societies and economies will be fought over. The chapter concludes with a summary of the key points made in these sections and an acknowledgement of some of the wider governance challenges that future citizens and policy-makers will face.

## The post-political city of the future and the corporatisation of London's public life

There is a real possibility that existing trends will continue into the future, and that over the next fifty years London will become an archetypal post-political city characterised by elite domination, social polarisation and divisive development. Since the early 1990s, city planning has essentially been concerned with the management of growth, and the sustainability of globally successful industries. Spatial policy has been used to try and ensure that some of the benefits of expansion reach poorer communities and neighbourhoods (Imrie et al, 2009). This dominant consensus has limited the articulation of meaningful *alternatives* and discussions over the type of city that London's citizens really want, and/or how these visions could be brought to fruition. In Ranciére's (2003) terms, it has been used to stifle formal political debate so that 'whatever your personal commitments, interests, and values may be, you perceive the same things, you give them the same name'. The result is that 'the only point of contest lies on what has to be done as a response to a given situation…so that [there] is a dismissal of politics as a polemical configuration of the common world' (paragraph 6).

The limitations of this way of thinking for social and economic justice have been evident, for at the same time as these post-political agendas of global 'success' have been rolled out, inequalities and social tensions have been expanding at a rate not seen since the early days of the industrial revolution. During the period 1996-2007, overall growth in UK GDP was 37.4%, yet remarkably only 0.7% of the population saw a growth in their incomes equal to or exceeding this (Murphy, 2011). By 2003 the top 10% of wealth earners owned 71% of the wealth and this 'is likely to be seriously understated as a result of the considerable shift in wealth offshore' (p.101). Most of these elites reside in London, where the poorest 50% of the population have less than 5% of the city's wealth, whilst the richest 10% possess 40% of income wealth, 45% of property wealth, and 65% of financial wealth (MacInnes & Kenway 2009). Similar patterns are evident in terms of health inequalities, land ownership, access to housing and employment and a plethora of other socioeconomic measures. They constitute the backdrop to the serious riots of 2011 and rises in crime and insecurity across the city. Unless there are significant forms of intervention and/or the economic system changes in structural ways, then it is likely that polarisation will continue to expand at a growing rate.

One additional aspect of post-political governance is its corrosive influence on the democratic process. In the name of enhanced 'efficiency', or what Tony Blair referred to as a principle of 'what matters is what works', state resources and service provision have been handed over to private operators and international investors on a previously unimagined scale. Since the mid-1990s this has resulted in global corporations taking over an expanding proportion of London's welfare assets (Raco, 2012). New forms of elitism have been established in which powerful private sector investors benefit by 'securing irrevocable contractual claims over taxation revenues that they will manage henceforth in their own private companies which they claim will undertake the tasks of the state so much better than the state could do itself' (Murphy, 2011). For Colin Crouch (2011), governance is therefore becoming increasingly dominated by global corporations who have the resources to acquire contracts and colonise a greater share of government activity. Much of London's new welfare infrastructure, such as its hospitals, has been provided under long term and hugely expensive Private Finance Initiatives in the name of policy 'efficiency'. The city has received 45% of all capital spending on PFI projects, making it by far the biggest recipient of any city in the UK (Musson, 2008). This will have a disruptive impact over the coming decades, as local managers desperately try to balance long-term asset payments with short-term budgetary pressures and demands. Many of the PFI contracts will still be in existence in the 2040s.

So by 2062 London may have become a post-political city. If current trends are sustained, its global industries will continue to dominate its economy, its politics, and its social order. This will be underpinned by a dominant consensus in which discussions over alternative ways of governing will be tightly framed, and policy 'innovations' will be limited to a series of piecemeal interventions, designed to tackle only the most severe problems and pockets of deprivation. Opposition will be managed and controlled, as political decisions become removed from the remit of political debate, and transferred to private companies who govern behind opaque contracts to impose quiet and ruthless forms of management efficiency. The role of the Mayor and other political actors becomes one of an 'intelligent client', seeking to manage the activities of private operators, who effectively determine how state services should operate and for whom. The city will continue to witness the residualisation of social exclusion, poverty, and inequality as core political concerns. The existence of structural inequalities will be blamed on crude understandings of individual and community (ir)responsibility. Little or no effort will be made to change the fundamental character of the city's economic base which continues to act as a vehicle for transferring wealth and resources from the majority to the affluent, and increasingly mobile, minority.

### The rise of a neo-Keynesian London and the revival of urban government

The financial crisis may be the prelude to a very different form of development politics in London and beyond. The dominant assumptions that have characterised neoliberal and hyper-global visions may be breaking down under the weight of failure and the very real limits imposed by environmental constraints. Rather than generating a long-term revival of such models, the current moment offers new opportunities for the development of alternative agendas and ways of thinking to take centre stage. New narratives and descriptions of what future development should look like are beginning, albeit tentatively, to emerge, perhaps most explicitly in the writings of authors such as Tim Jackson (2011) with his recent manifesto for *Prosperity Without Growth*, or Ha-Joon Chang's (2011) alternative agendas for future economics. It is, therefore, possible to imagine that by 2062 London will have become a showcase for a *neo-Keynesian* revival with states once again taking a direct, interventionist role in shaping social and economic affairs.

This could take on a variety of forms. To begin with, the contradictions and costs associated with privatisation may have triggered a renewed desire for coordination and the re-creation of top-down forms of bureaucratic governance and management. Such an approach could, at the very least, see the implementation of programmes of nationalisation of key economic sectors and welfare services. Governance may be replaced by government, as states look to take greater control over the management of cities and populations in response to the rather chaotic and fragmented structures that have emerged in the wake of neoliberalisation. There may be a rediscovery of the public realm and lively discussions over what might constitute the public interest in London. As Marquand (2004), and others, have long argued, under neoliberalism it has become increasingly difficult to articulate a political agenda that promotes public intervention as a vehicle for the creation of better cities. A reformed politics may see the revival of such agendas and the empowerment of public professionals to take greater control over key assets and industries.

It would also provide a partial solution to the growing malaise that exists in London and elsewhere in the UK towards formal democratic and state-led institutions. Empowered public sector bureaucracies could provide an important focus for the mobilisation of political agendas, as politics would once again matter. It would 'make a difference' whether or not politicians of different views were elected, as their programmes would have the capacity to actually re-shape the city, re-order how it functions, and change the day-to-day lives of its citizens. This could act as a catalyst for democracy and arrest the 'post-political' erosion of public debate and political life that the city has suffered.

Neo-Keynesian principles would also change the relationships between communities within London, and between the city and the rest of the national economy. A re-invigorated planning system would involve a renewed emphasis on spatial justice, and seek to ensure that London's aggregate economic growth would be redistributed more evenly, both within and outside of the city. This would, of course, meet with stiff resistance from business groups, and others, for whom any attempt to restrict economic growth in the capital would undermine its longer-term resilience (Raco & Street, 2012). Others, however, have called for a de-centring of the UK economy and the cultural dominance of London and the South Amin et al, 2003; Imrie et al, 2009). The city is criticised for being parasitic, rather than propulsive, and any serious neo-Keynesian approach would look to find ways of increasing the contributions made by its financial service industries, and wealthier individuals, and use the welfare system to redistribute resources to areas and communities of need. Again, the implications of this for London could be enormous, and in large part depend on how the global economy evolves in the coming decades.

Moreover, in a neo-Keynesian London some of the problems associated with the city's past may re-assert themselves. The transfer of the city's assets to state bodies may not necessarily generate new forms of collective ownership. Traditionally in some policy fields, such as health care and housing, Keynesianism of the past led to the concentration of decision-making power in the hands of professionals and experts which, in some places, created distant, top-down forms of administration and decision-making. Vulnerable groups such as those with physical impairments and/or the elderly suffered particularly badly under monolithic welfare structures (Oliver, 1990). As Milewa et al (2008) have shown, post-war welfare organisations were often deliberately insulated from political criticisms and community-led protests. Neoliberalism emerged in a context where authors from the political left, as well as the right, were attacking the growth of state bureaucracies for their detached and opaque structures (Gough, 1979; Offe, 2003). An enormous amount of social research on power elites, and modernist planning practices, in the post-period also laid bare the sexist and often partisan nature of some public sector bureaucracies in the wake of nationalisation (Saunders, 1979). It may become more difficult for community and third-sector groups to get their views heard. New forms of political closure could be enacted, based on paternalist forms of governance. If poorly managed, then many of the gains made through community activism in recent decades could be reversed.

## A London of shared and common value

In this final section, we imagine that a third scenario emerges based on a reformatting of capitalism. It emerges through a new urban political economy that takes place as a result of global struggles and conflicts in which London, and other global cities, provide battlegrounds for the articulation of different future visions between political, commercial and civil society interests. However this struggle does not end in complete or simple victory for either capital or labour. London in 2062 has been created via a new form of consensus capitalism: one which focuses on the relationship between societal and economic progress to create growth. In 2012 this was best represented by the concept and principles of 'Shared Value' (Porter & Kramer, 2011), and in this section we discuss what the implications of what such an outcome might lead to.

The ability to accept and connect with 'the other' when combined with the will to succeed is what has shaped the city over centuries and is what created the London of 2062. How these attributes impact on the behaviour and interaction of the political and governance, commercial and civil society institutions and the actors who make and remake the city, shapes London's destiny. In turn, it also shapes other cities, hardwired into the global city network. As Porter stated in 2012 'at a very basic level the competitiveness of a company and the health of the communities around it are closely intertwined' (Porter & Kramer, 2011). In 2062 the ability of Londoners to take a risk, and to accept and connect to others, no matter their background or origin, has resulted in a new institutional framework for commerce and civil society, where economic competitiveness and social progress are not just intertwined but interdependent. A new political economy of shared value has been born.

We imagine that a London of shared value in 2062 is really a product of the global uprisings that started with the Arab Spring of 2011, and the Global City Uprisings that will emerge in the coming decades. It is not difficult to imagine that the ever increasing connectivity London shares with other key cities across the world, the widespread use of technology and social media and the production of the best educated young urban generation (via the investment in education between 2000 and 2010), will give rise to a new movement that eventually reshapes, not only civil society, but governance and commerce structures. London will become a crucial node in a network of 'City Uprising' social movements, made up of well-educated consumers as well as the disenfranchised and disconnected. These groups were prepared to risk challenging the oligarchy (the so-called 'feral elite') previously created between commercial and governance interests, driven by the realisation that the levels of inequality and unsustainable urban living were a direct product of the urban political economies, operating in global cities, controlled by these interests. The City Uprising movements reformed civil society in London as well as other global cities, establishing a broad set of common values centred around trust, transparency, social solidarity, moral purpose and civic action, aimed at reclaiming some of the processes by which 'their' city was produced and reproduced. Global connectivity allowed individual actors sharing these values and behaviours to become linked, as previously global capital had become, creating a strong and distributed civil network, capable of challenging the institutional status quo, whilst remaining rooted in the communities they grew out of. The impact resulted in civil unrest and struggle for many years as London, and other global cities, convulsed in the wake of huge economic instability. Political, commercial and governance institutions were destroyed and remade in response. However, the result by 2062 was not what was predicted.

As the financial sector's political and economic power and influence dwindled, London's global technology, green manufacturing and utility and service industries took centre stage. Clearly distancing themselves from the socially unacceptable commercial behaviours of 'big business' (seeing the competitive advantage in doing so), and so new, or emergent, as not to be tied to the previous discredited oligarchy, a new commercial institutional framework evolved in London and other global cities. It represented a reformatting of capitalism, and its relationship with the consumers it depended on and the communities it was based within. These approaches began to pursue economic competitiveness and social progress at the same time. This new institutional framework enabled commercial interests to seek new markets in the developed, as well as undeveloped, world, seeking competitive advantage and increased market share through the adoption of shared value approaches. This new breed of global commercial operation created new institutional approaches, seeking to create additional economic *and* social value, by connecting more closely with the communities they were situated within as well as their customers; developing local supply chains, to reduce costs and build resilience into production; creating new relationships with local consumers, to create new products and enter new markets; and reengineering their value chains, exploring new ways of delivering, including increasing local social investments and developing new forms of private/social ownership. This new set of priorities, linked to their profitability, created new institutional links between commercial and civil society interests built on shared moral, ethical and civic values as well as the creation of shared economic and social value. This new form of capitalism not only embraced the civic and moral values of City Uprising, it inculcated it into its own DNA.

These new institutional models, created in London through the disruptive and anarchic convulsions of the 2020s, were adapted and adopted throughout the major global cities, where the corresponding commercial and civil society institutions – now more interconnected than ever before – responded to the structuring forces unleashed in London and other major cities by creating similar forms of business model and ownership, fusing commercial interests and the civic interests of City Uprising. This radically changed the processes of the urban political economy, through the dynamic tension between commercial, governance and civil society interests.

This reformation of capitalism resulted in the reformation of the urban political process, and vice versa. London's ability to foster 'risk takers' and 'chancers', coupled with its incoherent informality, created the conditions for this dramatic reformation. City Uprising presented the touchstone for those willing to take a chance and challenge the neoliberal elite, just as deregulation and privatisation had offered the same chance to commercial interests to challenge the consensus in the 1980s. This resulted in a realignment of governance, commercial and civil society institutional frameworks, increasingly driven by a shared vision of civic culture, common purpose and social solidarity. London in 2062 has been created via a new 'consensus capitalism' - one which focuses on the relationship between societal and economic progress to create growth. The line between civil society and commerce has become blurred, creating new institutions, organisations and arrangements focussed on shared moral, ethical and civic values as well as the creation of shared economic and social value.

## Conclusions

Whatever course the global economy takes over the coming decades, it is clear that London will be faced with a number of competing challenges for control of the city. In this chapter we have provided a lens through which to imagine these changes, set out some of the political economic possibilities that may emerge and some potential directions of change. We have argued that as a premier global city, London will be at the receiving end of changes in the global economy, but will also act as a centre for change and influence well beyond its borders (Massey, 2007). There are other factors that we have only touched on, including environmental change, the city's demographic future, and technological innovations. The wider political environment is also subject to enormous uncertainties, and while we have tried to adopt an optimistic tone in the discussion here, there are other trends in existence that raise altogether more worrying future prospects. For example, parties of the Far Right are becoming increasingly powerful across the European Union (EU), and some of the assumptions over the free movement of people and trade that have underpinned London's economy and changing social make-up may be a product of their time. Recent discussions in the EU over the 'reception capacities' of places, and the use of borders to manage migration flows, represents one possible future that will have significant implications for London Raco & Tasan-Kok, 2010).

Notwithstanding these wider changes, we have also argued that institutions matter, and that it is essential that government bodies and civil society organisations have the capacities and resources to implement programmes of action. This means strengthening existing structures where they work effectively and reforming and creating new ones when required. We disagree with authors, such as Beck (2008), for whom territorial state-systems have become 'zombie categories', with little substance in a cosmopolitan, post-national world. Institutions still act as a focus for political demands, and in our discussion here, we have outlined ways in which they can be reformed to change the form and character of governance in the city. In outlining some of Porter and Kramer's (2011) ideas on shared value, we have suggested what an alternative form of capitalism might look in future, brought about by political, economic, and social upheavals. Current models, we argue, are unsustainable, although it is possible to imagine the emergence of a divided, post-political city in which a tiny minority govern in their own interests. A return to neo-Keynesianism is, perhaps, unlikely in a context of austerity governance, but it is possible that the failure of existing systems might precipitate a return to earlier modes of government intervention, and renewed concern with direct social and spatial redistribution. Whatever path the struggle for the city takes, it is likely that it will continue to be influenced by the intangible attributes of Londoners described by MacInnes (1962); the ability of the city to nurture and bring together risk-takers who are prepared to take radical action that continually makes and remakes the institutions that shape the city. London has long been a city of extremes, with a cultural and symbolic significance all of its own. History shows that what happens in the city in the next fifty years will have a resonance long into the future.

# References

Amin A, Massey D. and Thrift N. 2003. *De-centering the Nation. A Radical Approach to Regional Inequality.* London: Catalyst

Beck U. 2008. *World at Risk.* Cambridge: Polity Press

Chang H-J. 2011. *23 Things They Don't Tell You About Capitalism.* London: Penguin

Crouch C. 2011. *The Strange Non-death of Neo-liberalism.* Cambridge: Polity Press

Gough I. 1979. *The Political Economy of the Welfare State.* Basingstoke: Palgrave, MacMillan

Imrie R, Lees L. and Raco M. 2009. London's regeneration. In: Imrie R, Lees L. and Raco M, editors. *Regenerating London: Governance, Sustainability, and Community in a Global City.* London: Routledge. p. 3-23.

Jackson T. 2011. *Prosperity Without Growth.* London: Earthscan

MacInnes C. 1962. *London: City of Any Dream.* London: Thames & Hudson

MacInnes T. and Kenway P. 2009. *London's Poverty Profile.* London: New Policy Institute

Marquand D. 2004. *Decline of the Public.* Cambridge: Polity Press

Massey D. 2007. *World City.* Cambridge: Polity Press

Milewa T, Buxton M. and Hanney S. 2008. Lay involvement in the public funding of medical research: expertise and counter-expertise in empirical and analytical perspective. *Critical Public Health.* 18: 357-366.

Murphy R. 2011. *The Courageous State – Rethinking Economics, Society, and the Role of Government.* London: Searching Finance Ltd

Musson S. 2008. Public-Private Partnership and the Geography of the State. Presented: Association of American Geographers Annual Meeting 15th–19th April 2008. Boston

Offe C. 2003. *The Contradictions of the Welfare State.* Cambridge: Polity Press

Oliver M. 1990. *The Politics of Disablement.* London: MacMillan

Porter M. and Kramer M. 2011 (January-February). Creating Shared Value. Harvard Business Review. Available from: http://www.hks.harvard.edu/m-rcbg/fellows/N_Lovegrove_Study_Group/Session_1/Michael_Porter_Creating_Shared_Value.pdf. [Accessed 07 August 2013]

Raco M. and Tasan-Kok T. 2010. Competitiveness, cohesion, and the credit crunch: Reflections on the sustainability of urban policy. In: De Boyser K, Dewilde C, Dierckx D. and Friedrichs J, editors. *Between the Spatial and the Social.* Hants: Ashgate Publishers

Raco M. 2012. The New Contractualism, the Privatisation of the Welfare State, and the Barriers to Open Source Planning. *Planning, Practice, and Research.* In press

Rancière J. 2003. Comment and Responses. *Theory & Event.* 6(4): 1-28

Raco M. and Street E. 2012. Resilience Planning, Economic Change and the Politics of Post-recession Development in London and Hong Kong. *Urban Studies.* 49(5): 1065-1088.

Saunders P. 1979. *Urban Politics – A Sociological Approach.* London: Penguin

# Let's sing, not shop: an economist dreams of a sustainable city

## David Fell

When imagining a London of 2062, it is easy to get excited about the possibility of personalised jet packs, hover cars and low-cost space travel. My personal hopes are just as fantastical: I want to see microwave oven-sized waste disposal machines, which generate energy and heat as a by-product, in every home; and 3-D printers that use nano-materials to build the products I want at the touch of a button.

As these imaginings immediately illustrate, there are dangers in looking far ahead: we may simply be left looking absurd, either now or in the future, once it arrives.

Nevertheless, there are uses for long-range scenario-planning. Firstly, it can help us to make normative judgments – do we prefer this possible future or that one? And, secondly, by navigating back from our preferred future, we can begin to think about the kinds of things that we might do now, or soon, that would steer us in the direction of our preferences.

The idea that we can choose our direction of travel is not merely an expression of belief in the power of democracy. It is also an assertion about the nature of an economy. An economy is not a closed system that tends towards dynamic equilibrium, the parameters of which are deterministic: it is, rather, a complex, adaptive system, the rules of which are socially determined and contingent. The economy is a human construct and, as such, we have the ability – and, indeed, the responsibility – to shape it to suit our needs. Over a period as long as fifty years, we should certainly be able to make the kinds of choices that will steer our economy in one direction rather than another.

The 'steering' is not, however, some mechanical act, in which we pull various levers and pulleys to make the 'machine' go in the direction we want. It is, rather, a more subtle and ultimately powerful process in which underlying behavioural norms are challenged and modified in a far more organic fashion.

I want to suggest that there is available to us a much more sustainable London in 2062, and a much less sustainable one; and which one we end up with will be, in large part, a function of the way in which a variety of norms pan out and interact. Here are three examples:

**Masculine/feminine** – London is presently a macho city, characterised by needlessly tall buildings, aggressive corporate behaviour, narcissistic decision-making and damagingly ruthless individualism. Left unchecked, these behaviours will continue to generate extreme levels of social inequality, the unrestrained

**How to cite this book chapter:**
Fell, D. 2013. Let's sing, not shop: an economist dreams of a sustainable city. In: Bell, S and Paskins, J. (eds.) *Imagining the Future City: London 2062*. Pp. 119-121. London: Ubiquity Press. DOI: http://dx.doi. org/10.5334/bag.p

consumption of finite physical resources and an environment of profound psychological stress for the majority of London's citizens. A more feminised city – attending to notions of care, concern, inclusion, small-scale production and consumption – would, by contrast, inherently counter such trends. A more sustainable London in 2062 would come about not through direct measures to, say, reduce $CO_2$ emissions, but instead, indirectly and more powerfully, through the development of a greater ethic of care.

**Walled/open** – a great deal of London's economic life currently happens behind walls. Corporate decision-making is opaque: wealthy citizens immunise themselves from their 'neighbours' by living in gated communities; political processes are dominated by lobbyists and careerists conversing in inaccessible settings. A London of 2062 in which these barriers persist would probably function as a city, but it could not possibly be described as sustainable. A sustainable London would be one in which inclusion and participation was ordinary, in which openness and transparency were normal. In this more open London, social injustices, environmental harms and wealth inequalities would be more apparent to all, increasing both the demand for change, and the political will to act. Improved outcomes would emerge organically from the change in the underlying logic of social interaction and would not need to be 'engineered' through interventions from 'the top'.

**Material/de-material** – the London of 2012 remains a citadel to consumer-led capitalism, even in the teeth of recession. Londoners, and the tens of thousands of tourists that visit the city, go shopping as if the world is going to end (!) and spend stupendous amounts of money on largely pointless products. It is conceivable that this could continue and that a London of 2062 will be wealthy enough to protect itself from the reality that will by then have come about, in which the effects of climate change will have become severe and in which a great many natural resources are either seriously depleted or have already vanished. But better, surely, to begin the process of weaning ourselves off our addictions, and to de-materialise our economy and our lifestyles. Let's learn rather than spend; let's sing rather than shop; let's stop with all the stuff.

At a time when unemployment in London – and, indeed, the rest of the UK and across Europe – is so high, and when governments and politicians are frantically seeking the economic growth that will save us from the present debt crisis, there is a risk of appearing somewhat disingenuous when offering suggestions for the short term that do not appear immediately to address the urgent problems faced by so many fellow citizens.

But it is at precisely such a time that the 'new' is required. It would surely be a mistake of the worst kind to spend inordinate efforts to return to some mythologised 'business as usual', for the sake of short term credit, when it is so obvious – to everyone? – that it was 'business as usual' that got us into this mess. In such a spirit, and in light of the three longer term themes just discussed, I offer three propositions for immediate action that could, I believe, not only begin steering London in the direction of genuine sustainability but could also deliver some shorter term gains that would benefit us all.

Firstly, I would like to see the language of competition replaced by the language of collaboration. Individuals and communities naturally collaborate with one another, but the discourse of business and politics has become monopolised by notions of endless competition. We need to reclaim the discourse, and reshape the space within which we make our decisions.

Secondly, I'd like to see a dramatic increase in the extent to which social and economic assets are under the direct ownership and control of communities. This would help to de-couple the 'real' economy from the financial economy; and would give individuals and communities a much more direct stake in the future.

Thirdly, I'd like to see us attend to the notion of 'sustainable play'. Human beings are inherently creative, sociable animals, but this better side of our nature has, like so many other aspects of our lives, been appropriated by market capitalism. We need to claim it back and demonstrate (to ourselves, as much as anything) that we can interact, exchange and be fulfilled without reliance on a piece of branded equipment. We don't see many advertisements encouraging us to go for a walk, for example, for the simple reason that it is exceptionally difficult for a corporation to make money out of us if we're out and about doing nothing so complicated as having a stroll. But going for a walk could, from such a perspective, be the most radical thing you do all day. Go ahead: take that step.

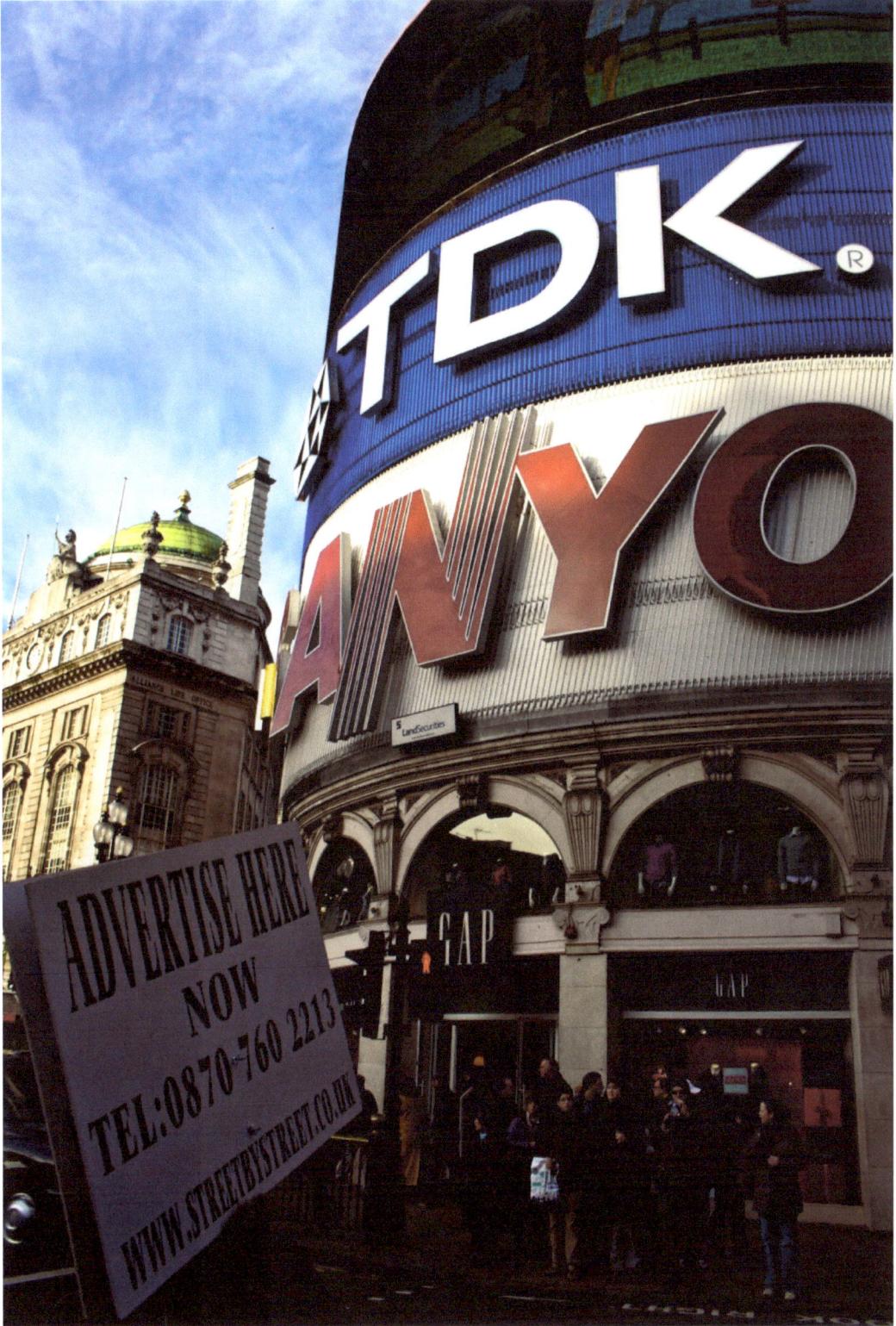

# Investing in futures

## Hannah Dalgleish

Any predictions about the composition of London's economy, or the kind of city it will be to live or work in, in 2062 would rely on wild estimates and presumptions. Looking back 50 years to the 1960s, when Wilson's Labour administration was in power and still presiding over a largely state controlled economy, who would have known we would be where we are now? A time when there were not too many global companies operating in London and the Welfare State and public sector as a whole was as strong as ever. Jobs were created and valued, and services delivered, for the collective good with minimal questions regarding efficiency or cost benefit analysis.

We are now in a period of continuing privatisation and scaling back of the public sector. A time when the state only intervenes when there is deemed to be a failure in the market. A time when the rise of the individual has led to much greater personal freedoms, which has further liberalised society at large, especially in London. The liberalisation of the economy has resulted in a surging increase in wealth and prosperity of a small minority of London residents whilst the rest have seen marginal improvements in real time earnings. More worrying is the widening of the gap between those at the top and those at the bottom, and the negative knock on effects that this rising income inequality brings. All of this is exacerbated in London, due to the concentration of financial services which has added value to commercial and residential rents and associated services. In the face of continued austerity, and prioritisation of the financial services, one can only dream that the London of 2062 could be so very different…

The London I would like to see in 2062 would, on the surface, look very much as it does now. However, if you looked a bit closer at the economic make-up of the city, and spent some time in its communities, you would soon begin to notice the differences.

After 45 years of investment in green technology, and state intervention and investment in other sectors, such as digital and extensive support of social enterprises and small businesses, the London economy has diversified and is far more robust. Investment in sustainable sectors has created a lot of jobs and training opportunities, especially for young people, and has increased innovation to the point that the UK (with London playing its part) is finally catching up with, and challenging, its continental partners in terms of commitment to greening the economy.

**How to cite this book chapter:**
Dalgleish, H. 2013. Investing in futures. In: Bell, S and Paskins, J. (eds.) *Imagining the Future City: London 2062.* Pp. 123-125. London: Ubiquity Press. DOI: http://dx.doi.org/10.5334/bag.q

The city is thriving, still seen to many as the economic powerhouse of the world, but it has lost its special privileges and influence. Governments are no longer bound to prioritise the interests of financial services. The Tobin Tax, brought in by the EU, has generated substantial income for all governments to reinvest back into public services and the increased corporation tax and income taxes of 80% of those earning more than £1 million a year hasn't driven a swathe of banks or private sector firms out of London. Nor has there been the widely expected departure to tax havens of those on higher incomes. It is now the widely held belief that markets can be a good servant but are always a bad master. This change in consciousness has brought about much tighter regulations of the banking sector and a slow move towards democratisation of private sector firms. Moreover, the rejection of GDP as the best way of calculating prosperity has also allowed for a faster move towards polices that capture and value social and sustainable outcomes. In a nutshell, growth is not the driving force behind decision-making in London anymore.

The state is around the same size as it was 40 years ago, but is now much more of an enabler with citizens taking a much more active role in delivering services. In London there has been a phenomenal rise in activism with communities running their local amenities, using their open space and taking ownership of key buildings and infrastructure. The outsourcing of services to Social Enterprises and the Voluntary and Community Sector has further helped to localise provision, and has also ensured vital services which derive no profit are still being delivered. The method of delivery has been transformed, with the collaboration and process valued as equally as the end result.

One of the most innovative progressions in London has been the move towards sharing the workload between a greater number of people. This has helped stimulate job creation and now more people than ever are able to work four-day weeks, enabling more time to be spent with friends, family and on general leisure. This transition has further helped the move towards more active citizenship with many people taking on voluntary roles within their own community.

The introduction of the living wage and a much fairer tax system has created an incentive to work. Jobs that once represented little in the way of value, self-worth or satisfaction are starting to become more appealing as people's attitudes in society are slowly changing, and gaps in pay aren't quite as staggering.

All of these national shifts are much more apparent in London which had the furthest to travel from a base of huge polarisation and income inequality. London is now a city that feels more cohesive, local communities are making decisions and the state is acting in those communities interest. House building is much more focussed on creating medium density family housing, aligned to the needs of growing population.

How did we get here? Or perhaps I should rephrase and ask how can we get here?

There is a growing collective consciousness around fairness. Over the last decade there has been a growing understanding of the need to take climate change more seriously, and to challenge the growth agenda. The election of Françoise Hollande to the French Presidency in 2012 suggests that moving towards a social democratic model isn't beyond possibility. All over London new creative bottom-up approaches are being developed and delivered, many of which can be viewed as pilots of best practice. The challenge is to channel the investment into these social projects and away from top-down public/private ventures, which see little benefit for end users and surrounding neighbourhoods.

London 2062 could be magnificent, a place where everyone is able to live, flourish and realise his or her potential. A lot of work needs to be done and we need to continue to challenge the status quo. In times of austerity creativity thrives, as does alternative thinking. We have a right to dream and the ability to make these dreams reality.

# Singing the helplessness blues

## Simon Cavanagh

The UK pop charts have played a part in chronicling, even forecasting the effects of recession on its population – The Specials' Ghost Town (Dammer, 1981) was the number one record a week before the last significant British riots, in the summer of 1981.

It appears to me, having to consider what London's economy will look like in 2062, the question this proposition poses is eloquently captured in Seattle's Fleet Foxes' 'new normal' anthem, Helplessness Blues (Fleet Foxes, 2011), but more of that later.

As London bumps along the bottom of a recession, it is very difficult to forecast ten years ahead, let alone fifty. It takes a leap of imagination to see how current or prospective Londoners will be able to put down roots in its clay and how they will contribute economically.

Working to deliver affordable housing in London, I am more aware than most of the many existing and prospective tenants' fragile grip on their communities, now described by many as exclusive or privileged locations. It would appear that Central London has no need for key workers or an increasingly marginalised working class, both of whom have played a major part in defining this city's economic past.

The globalised technology and financial services industries, held up as some of the keys to unlock London's economic future, will require educated and skilled workers. I doubt many of the next generation that will make up the workforce in 2062 will come from families in temporary or severely over-crowded homes. Adults and children need space and security to thrive, not a permanent state of rootlessness.

Many of London's core industries that had been their preserve are either gone, re-located or, as with the construction industry and its apprenticeships, more than ever focused on providing properties that attract buyers for their use as investment safe-havens. Less common are the hundreds of thousands of homes successive London Mayors have promised to support. There are some but not enough.

We have been here before, and so has London. The city has been shaped by fires, blitzes and economic downturns in its 2,000 year history, and all classes were present and correct then. What is a housing crisis compared to plague?

**How to cite this book chapter:**
Cavanagh, S. 2013. Singing the helplessness blues. In: Bell, S and Paskins, J. (eds.) *Imagining the Future City: London 2062*. Pp. 127-129. London: Ubiquity Press. DOI: http://dx.doi.org/10.5334/bag.r

So, either we imagine London 2062 as a gated community, populated by hedge-fund managers who take it in turn to serve school dinners and turn down hotel beds, or we keep calm and try and reflect.

We have to trust that the foundations of London's economic future will be laid in a re-instated Academy building programme; schools that focus and prepare young Londoners for a more competitive and globalised workplace. Just one graduate, Adele Adkins, has helped to re-energise a once moribund British music industry through a small independent record label. We will need more of her.

We have to trust that London's economy will be fuelled in part by the design, if not the manufacture of actual goods that can compete on a world stage with the best of the BRICS economies. We have to trust that these products will not all be missiles and smart gadgets, designed for fashionable obsolescence or oblivion. Goods that people need and want: Roberts radios, Dyson vacuum cleaners; industries that maximise profit through graft, skill and good design rather than smoke, mirrors and speculation. We will need more of those.

We have to trust that the wages will be decent enough, and the jobs fulfilling enough, to have supported the workless back into London's economy. We have to trust that they will be respected and valued for what they do and not what they earn.

We have to trust that Londoners will want to commute by better and more affordable public transport built and master-planned by governments that recognise the benefits of long-term investment in local infrastructure, not just the stuff that connects us to Europe and the world.

We have to trust that the homes Londoners return to, and want to invite their friends to, will be built to social housing standards in inner-city land banks and in a suburbia that has extended past a green belt that previously marked out the back gardens of a suspiciously high number of second homes.

We have to trust that all Londoners will once again be in a position to save for their future, but will also be more discretionary in their food-spending, so that the farms outside are not forced to sell their wares at a loss to a small number of wholesalers.

We have to trust that London stops for a moment and considers collaboration rather than competition with other post-industrial cities. In return, we have to trust that the governments of those other cities will recognise that London grew by offering a welcome, a possibility to succeed to everyone. That they will act accordingly, so that their citizens' best and brightest option isn't to have to leave. To be honest, London today feels like one of those places.

Many people have moved to London to make their fortune and to keep it close and that is their choice and their right but there are also many more that will come who, in the words of the Fleet Foxes, 'Would rather be a functioning cog in some great machinery serving something beyond them' (Fleet Foxes, 2011).

I think that line encapsulates the best of what it is to be a Londoner.

If London's economy isn't serving those aims in 2062, then it is likely that it will have had its day and it will be Hong Kong or Hamburg's turn. We can be more like Detroit with shades of Monte Carlo at its centre.

Many of us won't be around long enough to be proved right or wrong on our forecasts, but I hope we all spend the next few years laying some of the foundations to position this beautiful and cruel city somewhere between the boom and bust that have defined the previous fifty. Balance and the Fleet Foxes hold the key.

## References

Dammer J. 1981. Ghost Town. The Specials. 2 Tone: CHS TT17; Vinyl recording.
Fleet Foxes. 2011. Helplessness Blues. Bella Union: B004LQ19E0; Compact disc.

# Rethinking London's economy and economic future

## Myfanwy Taylor

If thinking about what London's economy might be like in 2062 seems fanciful, what about 2031? For this is the timeframe over which the present Mayor of London's Spatial Development Strategy (the 'London Plan') seeks to influence the city's development. Ideas about what will happen to London's economy in the future are thus already shaping Mayoral policies and priorities.

### Planning for growth

The vision of London's future in the London Plan (Mayor of London, 2011) is one of continued growth. It projects a future London of 8.82 million people (up 15.7 per cent from 2006) and 4.68 million jobs (up 16.5 per cent from 2007). Much of what follows is then aimed at providing the infrastructure and spaces to accommodate this projected growth: 'the only prudent course is to plan for continued growth' (p.23)

The targeting of growth is, in fact, open to challenge on many grounds, as was argued by Friends of the Earth (2010), Michael Edwards (n.d) and David Fell (n.d), amongst others, in their submissions to the examination of the draft London Plan. Here I focus on one specific challenge, namely that the growth that is imagined risks increasing inequality and worsening the present crisis of reproduction in London. This trend can be seen in development schemes, for example at Wards Corner, Tottenham (Wards Corner Community Coalition, n.d), that do not recognise and value but instead would displace and destroy existing economic activities which support livelihoods and deliver goods and services to London's diverse populations.

Interestingly, the London Plan (Mayor of London, 2011) includes no projections of the city's future inequalities. The solution to the problem is seen to lie in 'geographically targeted approaches to development and regeneration, focusing investment and action on places with the highest need' (p.25). Connections between the nature of economic growth and deprivation are not acknowledged, neither is the now overwhelming evidence that targeted regeneration displaces the poor and the businesses and organisations that serve them (Porter & Shaw, 2009).

**How to cite this book chapter:**
Taylor, M. 2013. Rethinking London's economy and economic future. In: Bell, S and Paskins, J. (eds.) *Imagining the Future City: London 2062.* Pp. 131-135. London: Ubiquity Press. DOI: http://dx.doi.org/10.5334/bag.s

## Growing the 'global' city

Part of the problem with the Greater London Authority's (GLA's) projected future economy of London is the nature of the growth imagined. 38% of the 776,000 new jobs anticipated by 2031 will come, the Plan projects, from business and financial services (Mayor of London, 2011). It is these sectors, then, whose interests the London Plan's policies primarily seek to serve. This form of representing the city and its economy as 'global' and 'world-class' makes the diversity of London's economy invisible, despite its importance to the city's long-term success and resilience (Buck et al, 2002). In other words, presenting London as a global city is a form of synecdoche, in which the part (financial and business services) is made to stand for the whole (London's diverse economy) (Amin & Graham, 1997). This matters because the presentation of London's success as resting fundamentally on the continued retention and attraction of international firms and workers in these sectors risks framing all other activities as secondary and unproductive of future growth. This reduces the space to put forward alternative approaches to urban economic development and planning that might offer greater possibilities for aligning social and economic development in London.

## Re-thinking London's economy and economic future

We urgently need to develop multiple alternative ways of thinking about London's economy and its economic future. The Greater London Authority's vision is irreconcilable both with the need to reduce carbon emissions and with a desire for a more just and equitable society. Furthermore, the choices London makes about what kind of economy it wants will have repercussions for people and places elsewhere, as indeed they have in the past. Where might we look for inspiration in developing alternative ideas for London's economy in 2031 or even 2062?

## The diverse economy

If we are to think imaginatively about London's economy today, we must first recognise that the idea of economy as being somehow separate from society – being subject instead to the self-regulating rationalities of an abstract market – is itself a construction of our own making (Mitchell, 2002). In this, J.K. Gibson-Graham's work is useful in showing that capitalism is but one way in which economic systems can be and have been organised by society (Gibson-Graham, 1996). Drawing on feminist and queer perspectives, she works to open up the notion of the diverse economy, 'representing and documenting the huge variety of economic transactions, labor practices and economic organizations that contribute to social well-being' (Gibson-Graham, 2008: p.615).

The diverse economy encompasses unpaid labour in the home or caring for others, informal exchange amongst friends, neighbours and members of alternative trading schemes, volunteering activities, as well as all kinds of ethical, green or otherwise-oriented market activities. As feminist geographers amongst others have shown, these activities are intimately connected with paid work and play an important role in keeping cities working. Seeing London's economy through this lens might open up a wider range of policies to support its diversity, such as stronger protection and support for trading and enterprise spaces for markets, community groups and small businesses, as well as provision of employment space close to residential space.

## Thinking about London from the global South

Postcolonial urban studies make a direct challenge to Western notions of modernity and development, suggesting that their relevance to cities outside their geographical origins is limited and, further, that that Western cities themselves might benefit from paying greater attention to theory and practice emerging from elsewhere (Robinson, 2002; 2006). For London, whose present global connections and influences rest on and remind us of its imperial and colonial past (Massey, 2007), postcolonial theory makes a particularly pertinent challenge.

Poorer cities may have much to teach London, amongst other cities, specifically in relation to the entwining of the social and the economic in cities. As Jenny Robinson argues (Robinson, 2006), because basic needs and developmental agendas are much more powerfully present in the global South, they more powerfully assert themselves into strategic debates regarding urban growth. This means 'local governments and development organisations are eager to find effective ways of expanding local economic opportunities through supporting the livelihood strategies of the poorest city dwellers or, at the very least, not disrupting existing activities' (p.153). Such a perspective would pay more attention to the skills and capabilities that the urban poor deploy in order to 'make do' in cities, as AbdouMaliq Simone (2001) has explored in relation to urban Africa, for example. Thinking about London from these departure points might prompt us to give greater emphasis to issues of urban reproduction and survival, recognising and valuing the spaces in which diverse economic activities are already underway and might be nurtured further.

## Already-existing alternative economies

Finally, we can take inspiration from the communities and organisations who are already working to protect and support diverse economies in London. In Brixton, an alternative currency has been launched to encourage money to 'stick' to Brixton and support local independent trade and business (The Brixton Pound, n.d). In Tottenham, the Wards Corner Community Coalition continues to fight against plans to demolish and re-develop Wards Corner, currently an important site for market traders, especially in speciality Latin American goods and services, and to develop an alternative Community Plan (Wards Corner Community Coalition, n.d) across London, we can look to the example of Just Space, an 'informal alliance of community groups, campaigns and concerned independent organisations that came together to challenge policies in the London Plan and take part in the "Examination in Public"' in 2006 (Just Space, n.d), which continues to play an important role in challenging the growth assumptions of the London Plan and in developing alternative perspectives on London's economy. Just Space's activities have been supported in various was by university-based researchers and students, many of them at University College London. Further sustained engagement between community groups and university-based researchers and students, amongst others, including through UCL's London 2062 initiative, offers great potential in the development and proliferation of diverse new ways of thinking about London's economy and economic future.

## Acknowledgements

In writing this paper, I have benefited from participating and tutoring in a course on 'Community Participation in City Strategies' as part of the MSc in Urban Studies at UCL, and from my involvement with the Brixton Pound, Wards Corner Community Coalition and Just Space. Thanks also to my supervisors, Jenny Robinson and Michael Edwards, and to the ESRC for supporting my PhD studies at UCL.

# References

Amin A. and Graham S. 1997. The Ordinary City. *Transactions of the Institute of British Geographers.* 22(4): 411-429

Buck N., Gordon I., Hall P., Harloe M. and Kleinman M. (eds.). 2002. *Working Capital: Life and Labour in Contemporary London.* London: Routledge

Edwards M. n.d. Draft Replacement London Plan / Economic Development Strategy: Comment and objection from Michael Edwards. Available from http://justspace2010.wordpress.com/welcome-to-just-space/submissions-about-the-plan/edwards-economy/. [Accessed 23 April 2012]

Fell D. n.d. David Fell – Economy. Available from http://justspace2010.wordpress.com/welcome-to-just-space/submissions-about-the-plan/david-fell-economy/. [Accessed 23 April 2012]

Friends of the Earth. 2010 (January). Friends of the earth (foe). Available from: http://justspace2010.wordpress.com/welcome-to-just-space/submissions-about-the-plan/friends-of-the-earth-foe/. [Accessed 23 April 2012]

Gibson-Graham JK. 1996. *The End of Capitalism (As We Knew It): A Feminist Critique of Political Economy.* Minneapolis: University of Minnesota Press

Gibson-Graham JK.2008. Diverse economies: performative practices for 'other worlds'. *Progress in Human Geography.* 32(5):613-632

Just Space. n.d About Just Space. Available online at http://justspace2010.wordpress.com/welcome-to-just-space/about-2/ [Accessed 23 April 2012]

Massey D. 2007. *World City.* Cambridge: Polity Press

Mayor of London. 2011. *The London Plan: Spatial Development Strategy for Greater London – July 2011.* London: GLA

Mitchell T. 2002. *Rule of experts: Egypt, techno-politics, modernity.* London: University of California Press

Porter L. and Shaw K. (eds). 2009. *Whose Urban Rennaissance? An international comparison of urban regeneration strategies.* London: Routledge

Robinson J. 2002. Global and World Cities: A View from Off the Map. *International Journal of Urban and Regional Research.* 26(3):531-554

Robinson J. 2006. *Ordinary cities: between modernity and development.* London: Routledge

Simone A. 2001. Straddling the Divides: Remaking Associational Life in the Informal African City. *International Journal of Urban and Regional Research.* 25(1):102-117.

The Brixton Pound. n.d. Key Facts. Available at http://brixtonpound.org/about/keyfacts/. [Accessed 23 April 2012]

Wards Corner Community Coalition. History. n.d. Available from http://wardscorner.wikispaces.com/History. [Accessed 23 April 2012]

# Housing, inequality and a property-owning democracy in London

## Michelle Hegarty

Over 30 years after Margaret Thatcher's radical reformist agenda set out to create a property-owning democracy in Britain, the majority of politicians, both on the left and right, have agreed that owning one's home is a good thing. The principles underlying this agenda have been to encourage pride, responsibility and independence amongst citizens. The political philosopher John Rawls considers that the aim of such a system is to realise the idea of society as a fair system of cooperation between citizens, who are regarded as free and equal. As the access to the ownership of property in London declines for those on middle and low incomes, what will the impact be upon these ideals, and in turn policy making, democracy and governance in London by 2062?

Home ownership in London is in decline. The National Housing Federation forecasts that in London the majority of people will rent by 2021, with the number of owner-occupiers falling to 44% by 2021. 2011Census data shows that the owner occupation rate in London has fallen from 58.9% in 2001 to 49.5% in 2011 (ONS, 2012). The National Housing Federation publication 'Home Truths 2012: London' shows that the average property in London costs £421,395 and the average annual London income is £26,962. This makes the average London house price almost 16 times the average London income (National Housing Federation, 2012).

The majority of politicians are keen to focus on the flaws in a housing market where property values are rising at an incomparable rate to earnings and leading leads to boom and bust cycles. But what politicians also recognise is the psychological importance to the British middle classes of the ownership of an asset that will appreciate over time, providing them with a pension, options for care in old age and a 'nest egg' for their children. Young people, the next generation, in London still aspire to home ownership, they aspire to escape poverty in old age and help their children – despite how unrealistic, or expensive, that may feel to them.

The policy response to spiralling house prices, from left and right leaning Governments, thus far has been to attempt to alleviate pricing pressures by promoting the supply of new housing. Despite this, the demand for home ownership in London continues to outstrip the pace of supply.

**How to cite this book chapter:**
Hegarty, M. 2013. Housing, inequality and a property-owning democracy in London. In: Bell, S and Paskins, J. (eds.) *Imagining the Future City: London 2062*. Pp. 137-138. London: Ubiquity Press. DOI: http://dx.doi.org/10.5334/bag.t

For the past 5 years, since the 'credit crunch' of 2007/2008, potential for a significant increase in house building in the near future has been seriously undermined. This is due to a lack of access to finance, for both developers and would be purchasers. While many will consider a more risk-averse approach to lending will go some way to slowing the pace of price rises, it does little to support new supply. In addition, the commitment to reducing the deficit from both main political parties suggests that any large-scale public sector investment in house building is unlikely any-time soon. The crisis in the financial markets has meant that securing finance to buy a home has become increasingly difficult. The impact has been felt strongly amongst those on low to moderate incomes who may be able to repay mortgages, but are unable to save the money needed for the high levels of deposit which lenders require.

Policies and programmes to assist first time home buyers are considered by many of those priced out of the housing market as unfair, as they can only help the chosen few. In fact some argue it only serves to keep prices inflated. Interference by the state in the housing market can be seen as the solution, or the problem, depending one's perspective. John Rawls suggests that in order for a property-owning democracy to thrive, institutions must, from the outset, put in the hands of citizens generally, and not only of a few, sufficient productive means for them to be fully cooperating members of a society on a footing of equality.

Recent history has suggested that few politicians are willing to contemplate policies that seek to reduce the prospect for the accumulation of wealth through property ownership, be that for homeowners or speculators and investors. Currently the majority of households in the UK are indeed homeowners and any such action may not be the most sensible thing to do for any politician seeking election to office.

But what happens when the balance of homeownership tips? The shifting picture of London from a majority of owner-occupiers to a minority, with the former locked out of the benefits of owning an asset, is likely to lead Londoners to look to their political leaders for a response on this issue, possibly ahead of other regions in the country. The policy response may be to encourage an increase in supply, reduction in demand through other housing options or to discourage the speculation on housing values in London.

Whatever the response needed, the more pertinent question may well be around the adequacy of the autonomy and power of London's political governance arrangements to respond effectively. The Mayor of London has recently expressed support for further measures of devolution to London recommended in a report (London Finance Commission, 2012) including all property taxes, including setting the rates of and being in control of all revenues from the full suite of property taxes. Such measures would enable London government further intervention in the housing market.

There are many who claim that the property-owning democracy ideal has achieved neither its social nor its financial goals, especially in London. Come 2062 will London be the first to see the death of the property-owning democracy or will politicians implement policy to stabilise the housing market.

## References

London Finance Commission. 2012. *Raising the Capital*. London: London Finance Commission

National Housing Federation. 2012. *Home Truths 2012: London*. London: National Housing Federation

ONS. 2012 (11 December). Key statistics for local authorities in England and Wales Available from http://www.ons.gov.uk/ons/rel/census/2011-census/key-statistics-for-local-authorities-in-england-and-wales/index.html. [Accessed 12 August 2013]

# Paranoia House

## Arthur Kay

*Stepped-up media reporting of violence and natural disasters have created a national ethos of paranoia and culture of fear.*

S Low

*News reporting capitalises on our greatest fears...focusing on symbolic substitutes rather than facing our moral insecurities and more systematic social problems.*

B Glassner

A vision of the London house in 2062. The only thing we have to fear is fear itself...

Terrence Fyed is a normal man living in the area of Nine Elms, South London. He has always kept up with world news and approaches associated problems pragmatically. A psychologist in 2012 would tell Terry that he suffers from crippling paranoia. However, in 2062 he is in good company, regarded by friends and family as remarkably blasé about world crises. The stories and images that have dominated world news for the last fifty years rule his life. Over time he has made a series of architectural interventions to his detached Victorian house, attempting to address specific fears derived from perceived local and global threats. The relocation of the Unite States embassy to the area, his house's proximity to the River Thames and the recently finished nuclear reactor at Battersea PowerStation don't help Terry stay calm!

Terrence's daily routine is dominated by the news. Whilst he does still go out to work and return home every day, the amount of panic he experiences is directly proportional to the amount of news he hears. Fear spreads virally throughout the 'global community', following the usual roots of apple-facebook global.

### Terrence's Architectural Retrofit

The key features or Terrence's retrofit are:

- In case of emergency flooding – space for underground water reserves.
- In case of emergency famine – live fish kept in the basements water reserves, tinned goods (10,000 stored in emergency escape route).

**How to cite this book chapter:**
Kay, A. 2013. Paranoia House. In: Bell, S and Paskins, J. (eds.) *Imagining the Future City: London 2062.* Pp. 139-140. London: Ubiquity Press. DOI: http://dx.doi.org/10.5334/bag.u

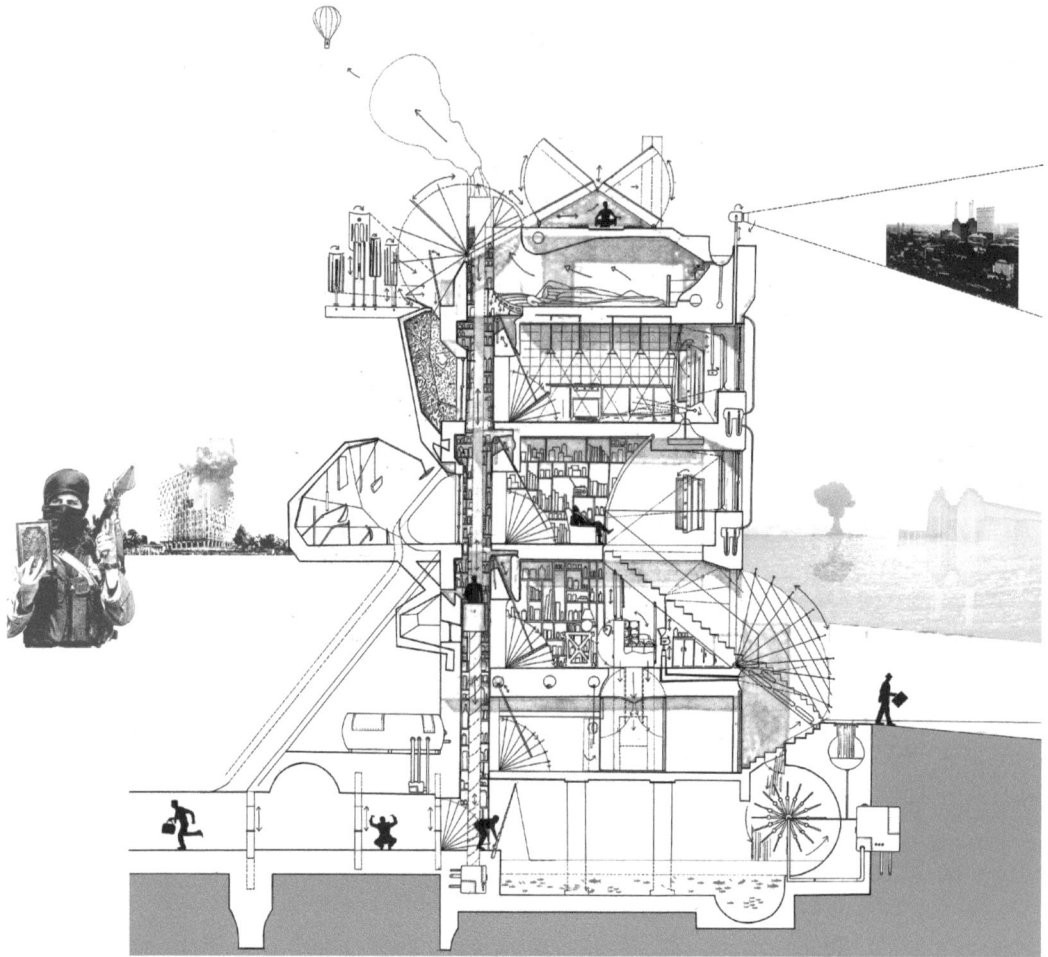

- In case of emergency unknown crisis – hydraulic lift and emergency aerial escape route via hot air balloon. This space acts as the backbone of the house providing access to the rooms and a physical escape, a vessel in which Terry can make good his escape over London away from any dangers.
- In case of emergency any-crisis – three air tight panic rooms, sealed to nuclear, chemical land ballistic terrorist attack.
- In case of emergency financial crisis – gold reserves if currency devalues.
- In case of emergency drought – water reserves in basement pool, decontamination areas for water and continual flow maintained through link to the water table.
- In case of emergency energy crisis – Terrence has ensured that hydroelectric, solar and wind energy sources are all included in his retrofit.
- In case of emergency mental distress – a series of idyllic dioramas have been assembled vertically by the hydraulic lift, providing a space for psychological release. These are in contrast to the world in which Terry lives, directly addressing some of the issues at hand.

Terrence has worked hard to retain the fabric of the building as he is anxious not to break any of the strict planning laws in the Greater London area, of which serious violation is now punishable by death. He has consequentially ensured that these alterations are not visible from the exterior of the building.

The house was our corner of the world, sheltering day-dreaming, our centre of intimacy. Now, in 2062, the idealised, proactive emotions of home have been encroached by increased globalisation, inevitable terrorist attack, environmental catastrophe, nuclear disaster, economic crises and political instability.

# Dreams

In the final section of the book we are invited to dream. In *Dreams* the authors range further than in previous chapters, and are less tied to our immediate concerns and expectations. These creative chapters take imaginative, speculative leaps into the future. They do not present certain futures, but new insights can be gained by following our authors' intuitions, hopes and fears to their logical, and sometimes fantastic, conclusions.

When considering the future, scenario planning is never far away. In *Dreams,* opinions of this technique's value differ. In one chapter, it is described as a structured way of combining current knowledge and predictions, balancing priorities and acting as an effective planning tool. Elsewhere it is characterised as an exercise in compromise, employing hordes of consultants, producing bland consensus statements. So, while this section sees a detailed evaluation of a number of scenarios and their implications for London, it also features visions of the future that dispense with the conservatism of the committee, and reflect individual imaginations.

The pieces are more than mere whimsy; these imaginative individuals are well versed in the literature, trends and challenges in their areas, so when they dream they are drawing on a deep and rich understanding of a topic. It may be that these chapters make connections and produce insights that would not arise easily through conscious, logical analysis. Each vision of the future provides a new vantage point from which to consider today's problems. We see the 'English question' from the point of view of the United States of Europe, take a bird's eye view of London's controlled flood zone and non-investment zones, and try to understand what has returned London to a state of nature. Time will tell if we do indeed become citizens of football clubs, rely on supercomputers to fight crime, or find ourselves wandering the scorched streets of a city with a forgotten past.

All of these possible futures rely on decisions that we take on the road to 2062. These dreams and nightmares can provide a stimulus, prompting us to work for, or against, the outcomes they describe.

The final message in *Dreams* is a warning, supplied by a poster that has somehow fallen back through time from the London Underground of 2062. It is clear that the tube is now a different, more dangerous, place.

# Scenarios and the future of London

Theodoros Semertzidis and James Paskins

## What are scenarios?

It is impossible to predict, with absolute certainty, what will happen tomorrow, let alone fifty years hence. Our lives are subject to myriad influences, physical and social forces that shape both the world we live in and the relationships between actors. In a city the size of London, the number of individuals and organisations is enormous, and the number of interactions is probably beyond calculation.

Thinking about the future in terms of current trends is a common approach. For instance, the introduction to this book includes predictions about global population, forecasts about how many of those people will live in London and projections about the implications for the capital's housing, transport and education sectors. These figures are important and thought provoking, they frame debate and help identify priorities. However, simply following trends does not allow for the possibility of substantive, disruptive, change (Lombardi et al, 2012).

When setting out policies that will determine the development of London's housing, flood defences, transport infrastructure or energy supply, it is not enough to say that the future is complex and unknowable. How can we deal with decisions made in an uncertain present that will have long-term consequences, extending into an even more uncertain future?

Even if we can't make accurate predictions or forecasts about the future, we can at least be clear about our assumptions, our decisions and their likely consequences. One way to do this is to consider different 'scenarios'. Scenario analysis has been defined as 'a tool for ordering one's perceptions about alternative future environments in which one's decisions might be played out. Alternatively: 'a set of organized ways for us to dream effectively about our own future' (Schwartz, 1991).

Scenarios are stories, descriptions of alternative destinations, a medium through which to share ideas about possible futures. The scenarios approach can embrace both scientific inquiry about

How to cite this book chapter:
Semertzidis, T and Paskins, J. 2013. Scenarios and the future of London. In: Bell, S and Paskins, J. (eds.) *Imagining the Future City: London 2062*. Pp. 143-153. London: Ubiquity Press. DOI: http://dx.doi.org/10.5334/bag.v

the future and real-world planning, bridging the gap between the two and helping to overcome cognitive biases (Xiang & Clarke, 2003).

Scenarios differ from forecasts and predictions, even a simple scenario can make clear the uncertainty that is present, but all too often ignored, when projecting a trendline. Future scenarios attempt to capture the richness of possibilities, stimulating decision makers to consider changes they would otherwise ignore. At the same time, they organize the possibilities into narratives that are easier to grasp than great volumes of data. Above all though they aim at challenging one's mind. When contemplating the future, it is useful to consider three classes of knowledge (Shoemaker, 1995):

- Things we know we know.
- Things we know we don't know.
- Things we don't know we don't know.

A well-drawn set of scenarios will pique interest and stimulate debate, helping people learn, re-perceive and reflect. One outcome should be a keener understanding of the assumptions about the way the world works, and hopefully a clearer view of the present. The end result is not an accurate representation of tomorrow, rather better decisions about the future. The point is not to predict the future, but liberate people's insights (Schwartz, 1991).

Pierre Wack's (1985) words sum up nicely what the purpose of scenarios is:

'Scenarios deal with two worlds, the world of facts and the world of perceptions. They explore for facts but they aim at perceptions inside the heads of decision makers. Their purpose is to gather and transform information of strategic significance into fresh perceptions. This transformation process is not trivial – more often than not it does not happen. When it works, it is a creative experience that generates a heartfelt 'Aha!' from your managers and leads to strategic insights beyond the mind's previous reach.'

## London 2062

The UCL London 2062 project considered the future of London in terms of resilience, wellbeing and sustainability (Bell & Tewdwr-Jones, 2012):

*Resilience* is the ability to recover from, adapt to and live with changes that are beyond our control. A city must have the capacity to bounce back from disruptions and this depends on emergency preparedness and response, government, citizens, economy and infrastructure. Resilience is enhanced when cities have diverse resources and systems, as well as networks of people and structures that can recover and adapt to changing conditions. These include energy systems, flood defence systems, water systems, food systems, waste-management systems and financial systems.

*Wellbeing* for the people of London is influenced by the city's social and economic conditions, as well the physical environment. Wellbeing concerns include security, health, air quality, culture and heritage. Maximising wellbeing can be considered an ideal for society, politics and the city.

*Sustainability* covers, among other areas, population, governance, housing and transport. The project recognised that there was a need for coherence in sustainability principles at all levels: across a city, a country and internationally. This is arguably more important than ever, due to complex inter and intra-dependencies that make any intervention problematic.

| Resilience | Wellbeing | Sustainability |
|---|---|---|
| Energy systems | Security | Population |
| Flood-defence systems | Health | Governance |
| Water systems | Air quality | Housing |
| Food systems | Culture | Transport |
| Waste-management systems | Heritage | |
| Financial systems | | |

**Table 1:** The fifteen drivers for change considered by the London 2062 project.

## Considering four scenarios

To illustrate the use of scenarios, and the contribution that they can make, we will consider scenarios developed by four different organisations. It is instructive to consider how, and why, scenarios differ. Different organisations will compile different future visions, reflecting their own concerns and focus. Differing definitions of key outcomes also play an important role. There will be also be differences in the drivers of change between organizations and groups that are devising future scenarios. The differences can be dictated by different uses of the scenarios, or how the outcomes are valued. Differences can also arise due to differences in geopolitical context, timescale, depth, or what the dominant force is considered to be.

The four different scenario sets are presented below, along with a brief summary and discussion. While there are no universally accepted definitions for sustainability, human wellbeing, or the resilience, the discussion of each set of scenarios includes a brief comparison against the fifteen drivers for change developed for the London 2062 project (Bell & Tewdwr-Jones, 2012):

### Arup scenarios

Arup is an independent firm of designers, planners, engineers, consultants and technical specialists. The Arup scenarios consider four visions of the UK in 2040: 'let it rip', 'technofix', 'carbon rationing' and 'fortress mentality'. The four futures are considered in terms planetary health and economic prosperity. Although the focus is the UK, worldwide effects are considered. The scenarios are broad but concise, providing a timeline and key facts and figures. This set of scenarios uses STEEP (Society Technology Economy Environment Politics) indicators. Most of the fifteen drivers of changed are covered, though nothing is mentioned about air quality and heritage, with flood defences, security and housing taking a secondary role in comparison with other drivers.

It is apparent that the economy and the environment are the main drivers and society is somewhat left out, although it is mentioned within each of the scenarios (Arup, 2009).

**Let it Rip** (positive economic growth, negative planetary health): Economic growth and consumerism have been pursued at the expense of the environment.

**Technofix** (positive economic growth, positive planetary health): Economic growth remains politically important, but development of green and innovative technology is promoted by state.

**Carbon rationing** (negative economic growth, positive planetary health): Carbon is the new currency and a strict and enforced scheme of carbon consumption imposed by the central UK government is affecting people's lives.

**Fortress mentality** (negative economic growth, negative planetary health): Energy poverty reflects economic poverty as people lose their jobs and homes.

## DHL scenarios

This set of scenarios comes from DHL (Deutsche Post AG), a multinational mail and logistics services group. They developed five scenarios covering the prospects of their business in 2050: 'untamed economy–impending collapse', 'mega-efficiency in mega cities', 'customized lifestyles', 'paralyzing protectionism' and 'global resilience–global adaptation'.

The futures take a global perspective, and are highly detailed. Although they concentrate on the logistics industry, they bring in a lot of detail about the world in general. Furthermore, it is clear that a great deal of time and effort has gone into compiling the scenarios, producing high quality results. Professional input was sought, including Peter Schwartz and Professor James Allen Dator, world leaders in future studies.

A thorough process was used to choose the key factors in this set, an initial list of 62 factors was reduced to fourteen. The key factors included those heavily linked to the logistics industry, and the main driving forces seem to be the economy, technology and trade. Most of the fifteen drivers of change are covered, however, health, food systems and particularly housing are not covered in detail, and flood systems and heritage are completely absent.

The scenarios were the result of a search for robust strategies to widen the company's perspective, continuing their 'Delivering Tomorrow' series. The study aims to foster dialogue about the future of logistics by describing a number of different pictures of the world in 2050 (DHL, 2012).

> **Untamed economy–impending collapse**: The world is characterized by unchecked materialism and consumption and quantitative growth is blooming while sustainable development is rejected.

> **Mega-efficiency in megacities**: Megacities are both the main drivers and beneficiaries of a paradigm shift towards green growth.

> **Customised lifestyles**: Individualization and personalized consumption are pervasive worldwide, due to increasing education levels, considerable technological progress and growing global affluence.

> **Paralyzing protectionism**: Globalization has been reversed triggered by economic hardship, excessive nationalism and protectionist barriers.

> **Global resilience–local adaptation**: A high level of consumption thanks to cheap, automated production initially characterizes the world.

## SLU scenarios

This set of scenarios comes from SLU (Sveriges LantbrukUniversitet), the Swedish University of Agricultural Sciences. There are five different futures in this set of scenarios: 'an overexploited world', 'a world in balance', 'changed balance of power', 'the world awakes' and 'a fragmented world'.

The horizon of the scenarios is 2050 and the perspective is both global and regional (Europe). The scenarios give a clear idea of the possible future, without going into very much detail. Social, political and environmental factors were the driving forces, with some coverage of food production and land use.

Eight main general factors were used for the global analysis and seven for the European analysis. The method used for the study is called general morphological analysis, which is applicable because several of the factors analysed are not quantitative. Most of the fifteen drivers of change are covered, but some are either mentioned without expansion, or not mentioned at all: air quality, flood-defence systems, waste-management systems, security, health, heritage and housing.

The scenarios have been constructed by researchers from different disciplines guided by the Swedish Defence Research Agency (FOI), and the scenarios were constructed as starting points for identifying challenges facing food production and land use, and as a basis to formulate research issues within the research programme 'Future agriculture – livestock, crops and land use' (Öborn et al, 2011).

**An overexploited world**: Population growth is high and poverty is prevalent in the world, while climate change is large and there is considerable pressure on land resources.

**A world in balance**: Economic development is strong and population increase is lower than the UN's forecast, while global warming is kept low and pressure on land is limited.

**Changed balance of power**: Population growth is relatively low, climate and the environment are not priorities and power has moved to the East.

**The world awakes**: Population growth is as the UN forecast and action is taken towards sustainable development

**A fragmented world**: Population growth is high, there are no measures to regulate climate change and pressure on land resources is very high.

*Natural England scenarios*

This set of scenarios comes from Natural England, a non-departmental public body of the UK government, responsible for ensuring the protection and improvement of England's natural environment. There are four different futures in this set of scenarios: 'connect for life', 'go for growth', 'keep it local' and 'succeed through science'.

The horizon of the scenarios is 2060 and although they concentrate on the UK, many of the points could be applicable elsewhere. The scenarios are the most detailed considered here, perhaps too detailed. They concentrate on the natural environment, but other aspects are fully analysed as well.

There are fourteen main drivers used, covering many important issues. Though the work concentrates on the natural environment, the drivers used are applicable to other research areas. Most of the fifteen drivers of change are covered in detail, apart from flood-defence systems, waste-management systems and air quality, which are not mentioned.

This work has been conducted to support Natural England's approach to strategic thinking and in particular, the development of its long-term vision for the natural environment. Exploring a range of plausible futures will help Natural England anticipate and appreciate some of the long-term challenges and opportunities facing the natural environment (Creedy et al, 2009).

**Connect for life**: People now connect through vast global networks, though decisions and economies are based locally.

**Go for growth**: Making money is a priority and economic growth continues to be driven by consumption and new technology. Few people worry about the environment and almost everyone continues to consume at will.

**Keep it local**: Society now revolves around nations feeding and providing for themselves. Resources are limited and are tightly controlled, but consumption remains high.

**Succeed through science**: The global economy continues to be driven by innovation and everyone relies on business to keep the country growing.

## Archetypes and London's futures

None of the eighteen futures above were devised with a specific city in mind. Can we apply these different visions of the future to London? It would be difficult, if not impossible, to prepare for eighteen different futures. It is, however, possible to identify common threads in the four scenario sets. In the section below, we have drawn out four archetypical futures and used them to categorise the scenarios.

The drivers of change in the scenarios described above are similar, and cover factors that are important for the survival and growth of a city. There are common narrative themes across scenario projects, part of this similarity is due to similarities in drivers. Despite these similarities, the stories themselves can vary considerably, due in part to the different emphases and immediate personal experiences brought by those constructing the stories. For instance, the media and the social environments bring different trends to our attention at different times. While there can be similar patterns, the devil is in the detail, since the impacts on specific decisions in specific environments will be defined by the detail most relevant to those decisions and environments (Natural England, 2009).

Many scenario exercises produce four different futures, with a few having one or two more or less. It is useful to consider these alternative futures as variations on a set of four archetypical alternative futures. It should be mentioned here that a worst-case or best-case scenario do not necessarily exist, since in every 'disaster' there are always 'winners' and 'losers', and utopias probably remain impossible dreams. All scenarios can be 'positive' to those who prefer them and negative to those who don't (Dator, 2009).

The first of our archetypes is continued growth or 'business as usual'. In this future, governments, educational systems and organizations aim to build a vibrant economy, and develop the people, institutions and technologies to keep the economy growing and changing, forever (Dator, 2009).

The second alternative future's common theme is social and/or environmental 'collapse'. In this future, the economy cannot keep growing in a finite world, and for different reasons that people fear collapse occurs. This collapse can come from invasion, hurricanes, tsunamis, earthquakes, rapid global warming, pandemics, and so on. The end result could be anything from a globalised New Dark Ages to extinction of all humans (Dator, 2009).

The third alternative scenario's common theme is discipline, a future that makes 'sustainability' the priority. In this future, continued economic growth is deemed undesirable or unsustainable, and people feel that precious places, processes and values are threatened or destroyed by allowing continuous economic growth. In this case, people wish to preserve or restore these places, processes and values. Also, others might feel that although continued economic growth is good, or necessary, given the extent of poverty in the world, nonetheless we live on a finite planet with rapidly depleting resources, and a burgeoning population and waste. Thus, survival and fair distribution are more important (Dator, 2009; Natural England, 2009).

The fourth alternative scenario's common theme is 'transformation through technology', or a paradigm shift future that overturns current assumptions about governance and economy, connected with a worldview and value shift, enabled by new technologies. In this future, technology is a transforming power through robotics, artificial intelligence, genetic engineering, nanotechnology, or even teleportation and space settlement (Dator, 2009; Natural England, 2009).

It should be noted that although futures may share a category they are not the same future. The categorisation is made because they share important characteristics. Some of the futures can be categorised with relative ease, as they match closely one of the paradigms. Others have been placed where they fit best, and in some cases, where the scenario has aspects of more than one paradigm, it is categorised as both.

| | | Business as usual | Collapse | Sustainability | Paradigm shift – Technology |
|---|---|---|---|---|---|
| **Arup** | Let it rip | ◆ | | | |
| | Technofix | | | | ◆ |
| | Carbon rationing | | | ◆ | |
| | Fortress mentality | | ◆ | | |
| **DHL** | Untamed economy - impending collapse | ◆ | ◆ | | |
| | Mega-efficiency in megacities | | | | ◆ |
| | Customised lifestyles | ◆ | | | |
| | Paralysing protectionism | | ◆ | | |
| | Global resilience - global adaptation | | ◆ | ◆ | |
| **SLU** | An overexploited world | ◆ | ◆ | | |
| | A world in balance | | | | ◆ |
| | Changed balance of power | ◆ | | | |
| | The world awakes | | | ◆ | |
| | A fragmented world | | ◆ | | |
| **Natural England** | Connect for life | | | ◆ | |
| | Go for growth | ◆ | | | |
| | Keep it local | | ◆ | | |
| | Succeed through science | | | | ◆ |

**Table 2:** The eighteen futures from the four scenarios under consideration can be classified using four archetypes.

One thing that becomes apparent straight away is that DHL's futures tend to be towards 'business as usual' and 'collapse'. SLU also has two futures that are 'business as usual', and all others have their futures dispersed across all four main categories. Taking a closer look at the table, we can see that seven scenarios are 'collapse', six are 'business as usual', four are 'paradigm shift-technology' and four are 'sustainability'. In two cases a 'business as usual' future was also a 'collapse' future, which seems to suggest that people believe that continuing on the current path of economic growth will result in collapse at some point. The path of a 'paradigm shift' assisted by 'technology' seems to be the best-case scenario, while a 'sustainability' future, where problems have reached a point where immediate changes in attitudes and discipline are needed to achieve survival in a sense, seems to be the least desirable case.

## London's futures

We will conclude by considering what the scenarios can tell us about the London we might expect in thirty to fifty years. We will also consider how the futures can be achieved or avoided. Again, it is important to notice that specific futures might be positive to some and negative to others depending on their interests, and especially when a city is the centre of attention.

### London in a 'business as usual' future

This London is in the grip of environmental insecurity, with a rise in temperature and modified weather patterns. Since the sea level has risen, massive expenditure for flood protection is required. There is increased demand for energy and raw materials, while at the same time natural resources are bcoming more expensive. Traffic has increased, and waste disposal and air pollution present massive problems. The pressure on land is considerable and the availability of clean water is low. There will probably be occasional climate shocks, requiring state intervention.

The population is growing and there are densification problems. The gap between rich and poor has widened, and there is increased demand for an overstretched health services. The average age of the population is increasing, and London is increasingly reliant on inward migration. The city's knowledge industry has not deteriorated yet, but is under threat due to lack of resources and a global power shift to the East.

London is still economically powerful, though increasing temperatures and flood risk are beginning to present problems. At the same time, widening social gaps, such as marked income inequalities, are leading to increased social unrest. In the longer term environmental degradation is expected to lead to an economic slowdown, with some predicting catastrophic collapse.

### London in a 'collapse' future

Lifestyles have become unsustainable, natural resources are scarce and there is a considerable rise in temperature and sea levels, leading to major flood risks for the capital. Waste disposal and air pollution are massive problems, especially since technological development is lagging. The massive pressure on land resources and limited clean water availability, are sparking social unrest. This insecurity has cause major health problems. The city's population hasn't drastically reduced, due to high inward migration.

London's citizens must deal with underfunded infrastructure, flood risks and overcrowding. Large social units cluster together in the inner city, and fortress gated communities are becoming more prevalent, fuelling intergroup aggression. The centre of London is very densely developed, while in the outskirts, development is uncontrolled and disorganized. Though the South East is not a centre of gravity anymore, London is still attractive as a dynamic metropolitan centre, since some funding has been used to maintain or modernize infrastructure. Overall the city's economy is contracting. A slow down in imports or exports, has been accompanied by entrepreneurs and professionals leaving the city. Many consider the city to be in terminal decline.

### London in a 'sustainability' future

Carbon is the new currency, and a strict and enforced scheme of carbon consumption imposed by the central UK government affects everyone. Energy prices are high and tariffs on emissions from fossil fuels limit their use. Temperature and sea levels have risen, but not radically. Some adaptation to climate change is necessary, and investment in mitigation measures needs to be maintained.

The city's population has grown, due to a combination of technology, medicine and immigration. Availability of clean water for such a large population presents problems, and water recycling schemes are being extended. There has been a restructuring around smaller localised service centres. Parks and green areas, along with roofs and walls of buildings are used for small-scale farming. Large corporations have declined, with smaller firms comprising a larger section of the city's economy. Overall, economic development in the West is weak, unlike in China and India, but the UK and especially London is growing modestly in economic terms.

*London in a 'paradigm shift - technology' future*

There are stringent environmental regulations for industries and everyday life. Carbon pricing is used for all products and services. The maximum warming target of 2°C has not been reached, limiting the negative impacts of climate change. Pressure on land resources is limited with urban agriculture, green retrofit, green roofs and walls becoming more common. Biospheres exist within the urban fringe, creating a looser urban fabric. Waste and wastefulness have been significantly reduced over the last generation. People trust technology to enable growth within environmental and resource limits, but some worry it may not always have the answer. The population has increased due to medicine, technology and immigration. More people have also moved to urban areas. London is a prime megacity, which is the main driver and beneficiary of a paradigm shift towards green growth.

The city remains the epicentre of social, economic and political development and provides an attractive lifestyle, while living quarters are well guarded. London is a technological centre of gravity and still driving economic activity as a world leader in green technology, though international financial sector is much reduced.

## Conclusion

Some, or none, of these things will come to pass. But the point of scenarios is not to provide a crystal ball, instead they ask well structuted 'what if?' questions. When done well, they do so in a plausible way, with an internal consistency that makes participants question their assumptions, and follow the consequences of interventions to their logical conclusions. When we consider London in terms of pre-exisitng scenarios considered above, we begin to reveal possible impacts in many important areas. Hopefully, doing so will prompt questions about our priorities, the possible imapcts of today's decisions and our capacity to realise a sustainable future for London.

## References

Arup. 2009. *Planning exercise: Future Scenarios UK 2040.* London: Arup

Bell S. and Tewdwr-Jones M. 2012. *The UCL London 2062 Project.* London: UCL Grand Challenge of Sustainable Cities. Available from: http://www.ucl.ac.uk/london-2062/documents/2062-19July-2011.pdf. [Accessed 12 August 2013]

Creedy, J., Doran, H., Duffield, S., George, N. & Kass, G. 2009. England natural environment in 2060 - issues, implications and scenarios. Natural England Research Report NERR031. Available from: http://publications.naturalengland.org.uk/publication/31030. [Accessed 12 August 2013]

Dator, J. A. 2009. Alternative Futures at the Manoa School. *Journal of Future Studies.* 14(2): 1-18

DHL (Deutsche Post AG). 2012. Delivering Tomorrow - Logistics 2050: A scenario study. Available from: http://www.dhl.com/content/dam/Local_Images/g0/aboutus/SpecialInterest/Logistics2050/szenario_study_logistics_2050.pdf. [Accessed 12 August 2013]

Lombardi D.R., Leach J.M. and Rogers, C.D.F. 2012. *Designing resilient cities: a guide to good practice.* Bracknell: IHS BRE Press

Natural England. 2009. Scenarios Compendium. Natural England Commissioned Report NECR031. Available from: http://publications.naturalengland.org.uk/file/70040. [Accessed 12 August 2013]

Öborn, I., Magnusson, U., Bengtsson, J., Vrede, K., Fahlbeck, E., Jensen, E.S., Westin, C., Jansson, T., Hedenus, F., Lindholm Schulz, H., Stenström, M., Jansson, B., Rydhmer, L. 2011. *Five*

*Scenarios for 2050 – Conditions for Agriculture and land use.* Uppsala: Swedish University of Agricultural Sciences

Schwartz P. 1991. *The art of the long view.* New York: Doubleday/Currency

Shoemaker P.J.H. 1995. Scenario Planning: A tool for strategic thinking. *Sloan Management Review.* 36(2): 2.

Wack P. 1985. Scenarios: Shooting the Rapids. *Harvard Business Review.* November-December:139-150

Xiang W.N. and Clarke K.C. 2003. The use of scenarios in land-use planning. *Environment and Planning B: Planning and Design.* 30: 885-909

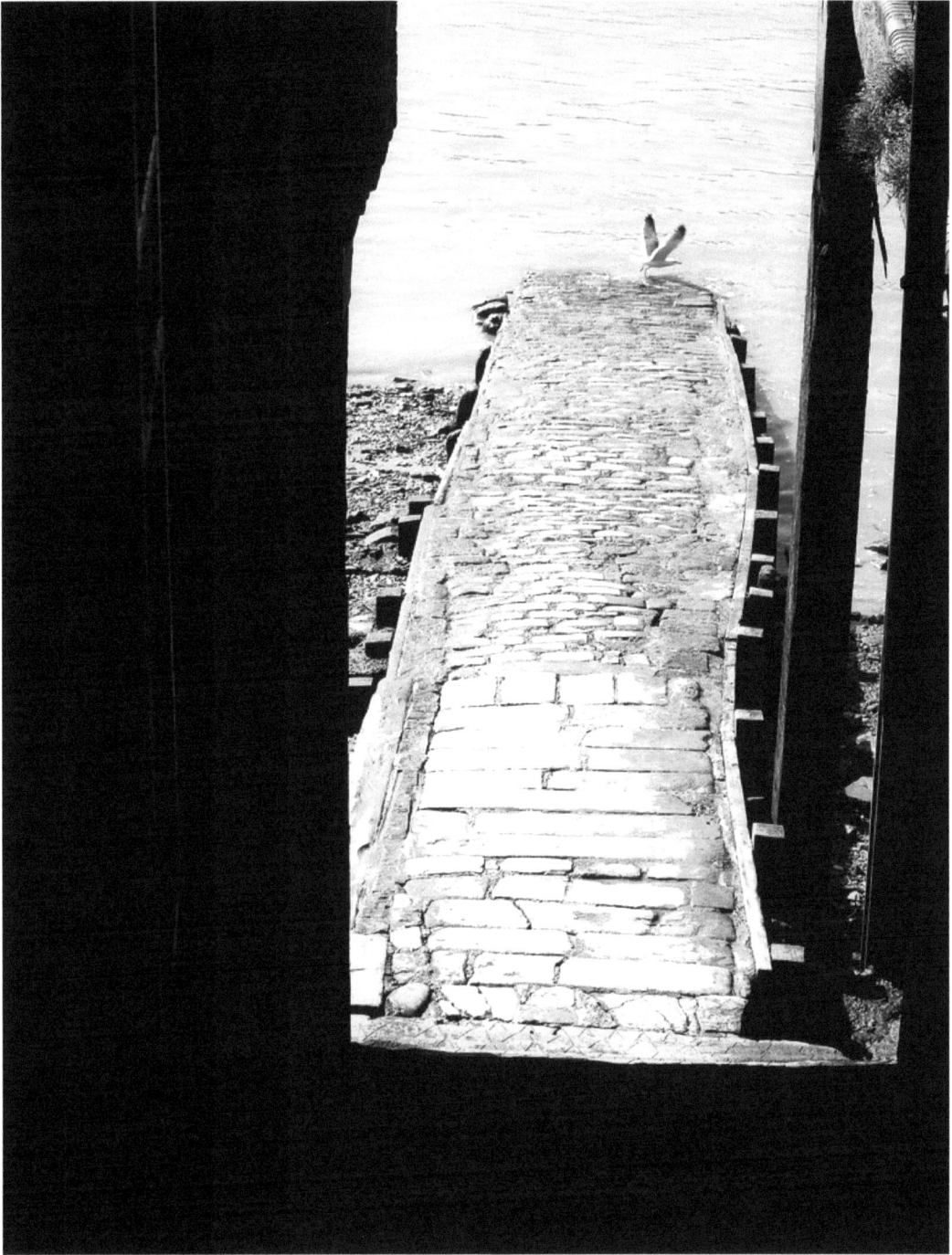

# No limits to imagining London's future

George Myerson and Yvonne Rydin

## Scenario development as a flawed compromise

Scenario development is now an established industry, employing hordes of consultants. This way of considering the future is typically an exercise in compromise. Workshops bring 'stakeholders' together to discuss drivers and barriers, opportunities and challenges. Post-it notes – square, round, hexagonal – get stuck to wall charts. Timelines become populated with views, and guesses, on what will happen when. Axes of key cleavages between different futures get proposed and, almost inevitably, four different futures emerge out of the intersections of two main axes. At each stage there is discussion and debate, but perhaps not always in-depth deliberation or examination. Differences of opinion may be resolved, or they may be fudged. Group think can often dominate specific sessions. The resulting narratives will, of course, reflect a wide variety of knowledge but not necessarily great depth. They are the outcome of many different voices being melded into more or less coherent story-lines.

The result is inevitably a compromise narrative; and even with careful crafting, tensions, discontinuities and contradictions remain. Involving multiple voices in creating a narrative means that certain points get included, because someone has a strong view about that point, or just has the loudest voice. The consultant is tasked with smoothing out the differences and creating coherence, but it requires particular skills in close reading and textual analysis to do this. Because we all read and write, these skills are often under-appreciated, in a way that quantitative skills, in a largely number-phobic society, are not. In addition, the starting point for much futures thinking is frequently technological change, again privileging a particular way of looking at the world. Understanding how governance and society might change can be comparatively neglected in terms of really thinking about possible futures.

So what would be an alternative approach? One possibility – that being taken by the EPSRC-funded CLUES project (UCL, 2013) – is to find a way to enhance the coherence of scenarios that

**How to cite this book chapter:**
Myerson, G and Rydin, Y. 2013. No limits to imagining London's future. In: Bell, S and Paskins, J. (eds.) *Imagining the Future City: London 2062*. Pp. 155-158. London: Ubiquity Press. DOI: http://dx.doi.org/10.5334/bag.w

have been developed through traditional workshop means. CLUES is using quantification as a means of teasing out the inherent inconsistencies in a set of existing scenarios to produce – it is argued – more robust and transparent scenarios. This has involved drawing on a wealth of published studies on likely future trends and validating this through discussion of quantitative tabular material within the research group and with external technical experts. Putting numbers to a narrative scenario poses questions which have to be answered with a degree of precision that really tests the scenario. The more bland statements of future trends are not accepted under this approach.

But a quite different approach is to dispense with the group means of scenario production, with its dependence on many diverse inputs from multiple experts, altogether. Instead, this approach involves relying on the individual imagination. This recognises the inherently artistic and creative nature of thinking about the future, and the need to root such an endeavour in the unfettered play of ideas, and trust to the coherence of a well-imagined story. The advantage of this approach is that is moves beyond the conservatism that is inherent in the workshop approach to scenario-building. It allows possibilities to be written that could not be said in a workshop, without nervous laughter and dismissal. It also dispenses with the sometimes spurious use of axes to structure scenario development, typically resulting in four such offerings. Each imagination is a sole contribution but there is no limit to how many can be developed. If the uncertain future is really being thought about, how much more realistic it is to consider one, three, six, twenty, fifty such offerings than be constrained by a matrix arrangement? In such offerings, all kinds of futures may be imagined. How many scenarios offered to policy makers to think through their options really consider the likelihood of a fascist state? How many consider radical new social values that would prioritise the views of children above those of adults? In the world of imagination, all futures are possible. So, in this spirit, here is one such imagining for London.

## London and the 5D football revolution

After the Great Devaluation of 2018 had failed to 'kick start' the Anglo-Welsh economy, individual regions were given responsibility for many of the functions previously discharged by the central state, including health, education and many aspects of social security. The London Metropolitan Region became officially bankrupt in 2029 when the region's authority, led by Consul Boris Johnson together with his Liberal and Labour partners in the London Triumvirate, was unable to negotiate further help from the European Regional Aid Fund. Yet strangely this was the beginning of a development that led to full-scale independence for the City State of Urbanitas, comprising the previous area of Greater London together with the Oxbridge Corridors and the Windsor and Eton enclave. How was such immense success possible from such a low point?

The first signs of what was to come pre-dated the financial and fiscal crisis of '29, occurring in the education sector of the city. It was in this field that the city's most successful global institutions began to take responsibility for the widening holes in the increasingly ragged state networks. The earliest examples were the Chelsea FC Charitable Academies and the Arsenal FC Popular Free Elementary Schools. Then, as the resources drained from the coffers of the regional authority, the capital's football clubs stepped in and started to fund more and more of the public institutions, starting with the Hotspur Health Trusts in north London and accelerating with the giant Gunners Housing Association developments around King's Cross and the Blues Riverside Estates, built to commemorate the fifteenth champions league win of 2031, the final flourish of the legendary reign of Sir Roberto Di Matteo. This more socially-oriented development supplemented the massive and profitable investment that had gone into hotels, restaurants and conference venues by the clubs.

As the global financial role of the City of London faded into the imperial past, the three giant football corporations, as they became, were the only truly global financial players left. Their prosperity was based on the expansion of the new 5D broadcast system, bringing the smell and touch, as well as the sight, of the match day to worldwide audiences of billions every week. The impetus for their foundational role in the establishment of Urbanitas began when the crowds started to decline in the aftermath of the great labour shake-out of '38, when the last financial trading companies left the now derelict Canary Wharf site, with the London Stock Market ceasing to trade the previous year. It became clear that without live crowds to sing and gesture and look picturesque, there was far less appeal to the new 5D match channel and the worldwide audience started to decline. In order to remedy this, the three football corporations began to pay the fans to attend—handing over vouchers at the turnstiles that could be exchanged in the clubs' networks of shops, schools, off licences, hospitals and other outlets.

The rise of the professional football fan as the region's main source of income was a rapid – and to economic and social experts, astonishingly widespread – development. By '44 – when the veteran Consul Johnson retired from public life to be replaced by his grand-daughter Livia – 88% of all employment in some areas of the city was in this new profession. The success of 5D Football after that cemented the football economy into place and underpinned the subsequent evolution of Urbanitas, with its vivid tripartite red, blue and white flag with the golden football in the centre.

The City State was now divided into the Red, Blue and White zones, each dominated by a massive football arena. These dwarfed the original arenas of the early 2000s. When the Emirates stadium was built for Arsenal football club in 2006 it was the biggest sports arena in the city; it was as if a huge alien spaceship had landed, visible from so many North London vantage points. That stadium became a holding space for people waiting for information about and access to the services that Arsenal provided. Islington Borough Council's offices down the road, that used to handle thousands of citizens' queries each day, had long been re-used to provide a free lunch canteen and social care facility for the disabled.

Governance was now firmly in the hands of the football corporations. Each of the zones had its own tribune, selected by popular acclaim at the last match of each season. Under the tribunate, there were officials responsible for functions such as waste disposal and road maintenance. Planning and development was dealt with by sub-tribunes, who presented pictures of large-scale developments on giant screens to receive acceptance by acclamation at the first match of every month. Massive lines of transportation by high speed rail and underground transit moved people from the edge of the city region to attend the matches that were the financial and cultural lifeblood of Urbanitas.

By contrast, the area south of the Thames had suffered the consequences of not having a successful Premier League club in the early 2000s. Massive population shifts across the river led to major dereliction, empty spaces and a take-over by wildlife; the north might be Red, Blue and White but the south was a mass of green, when seen from the air. And this could be readily seen, as planes came in to land at the three major airports located in south London and linked by rapid hoverfleet lines to the mega-arenas of the north. London remained a global hub, but that status rested on the interest of the world in the kick of a ball.

## References

UCL. 2013 (21 January). Challenging Lock-in through Urban Energy Systems. Available from: www.ucl.ac.uk/clues. [Accessed 7 August 2013]

# A city-state?

## Janice Morphet

The Swinging Sixties: a hundred years ago, London was at the centre of world news, with front covers of international magazines and the world's media curious about what was going on. London's experience of loosening the belt opened a period of creativity and change that started the separation of London from the rest of the country. Was this where the seeds of change, for the creation of the London city-state were sown? London was and wanted to be different.

The real turning point came in 1999. Devolution in Scotland and Wales and the powers given to the Mayor of London started an irrevocable process of separation. When Scotland voted for independence in 2014, the transition to the federal state of Great Britain began in earnest. Then, as now, it was the role of England that seemed to be the big issue and it is hard to look back without remembering the intense debates about the establishment of the English Parliament. The main concern was about where it would sit, and when Manchester was chosen, it seemed to galvanise London's position as an international, outward, global city and as a separate part of the UK. Even fifty years ago in 2012, the Mayor of London had more powers than any other elected politician apart from the Prime Minister.

Looking back, what led to the creation of the London city-state after the referendum in Scotland? And what difference has it made? Firstly, the UK referendum to leave the EU put London at odds with England, and it found that it had more in common with Scotland, Wales and Northern Ireland. The federal structure of Great Britain, as it was devolved, immediately demonstrated that the lack of leadership, vision and agenda for England was different from the drive and determination of London and its people. London was already separate in its governance; why not take the next step?

Manchester was always interested in running the rest of the country. When the Blair government set up Manchester as the second English growth pole after 1997, not many people noticed the way in which it was consistently privileged through government decisions made by both Labour and Coalition governments. Devolved spending, new local authority arrangements and eventually transferring taxation and civil servants to the Greater Manchester authority showed the Government's intent. When it was proposed in 2016 that the government of England should move to Manchester, London wanted the UK government to stay. This separation was needed to enforce

**How to cite this book chapter:**
Morphet, J. 2013. A city-state? In: Bell, S and Paskins, J. (eds.) *Imagining the Future City: London 2062.* Pp. 159-161. London: Ubiquity Press. DOI: http://dx.doi.org/10.5334/bag.x

some independence on England but London feared what might follow. In the end, the establish-ment of the English Parliament in Manchester and the associated move of some civil servants created a governance machine that was as small as that in the other devolved nations. Many civil servants, faced with a move to Manchester, opted out through retirement, particularly as future pensions could no longer be guaranteed.

The effects of this change in the seat of government for England could be anticipated at the time. However, what has proved more critical to London's position is the creation of the United States of Europe (USE) in 2057, one hundred years after the EU was formed in 1957. Since the UK/EU in/out referendum in 2017, the potential for differential relationships between the nations in the UK and the EU emerged. The decision of the UK to opt out and the subsequent decisions of Wales, Scotland and Northern Ireland to opt back in created a way for London to re-join the EU and sup-port its transition to the USE. This has been a difficult path to take, not least for London's financial economy and through the adoption of the Euro as its currency, but without this, London was faced with a major threat to its international position.

Yet despite these changes in government structures and institutions, is London any different from the way it was fifty years ago for ordinary Londoners? Firstly, there are more Londoners. London has continued to grow, not just in the centre but also in those high-priced housing areas of Barking, Romford and Dagenham that accompanied the East London airport expansions. Of course it would be difficult to accommodate so many people if there were still private cars, but the decision to abolish the use of cars in zones 1-2 in London has meant more buses and bikes, which have now become the predominant modes of transport. It is rare to see a petrol filling sta-tion or car park within these areas now, most having been redeveloped for housing in the 2020s and 2030s. It also has opened the streets to more walking and running which more people do on their way to work.

Much of London's housing looks the same, but the major housing retrofitting programme which began after the nuclear energy crisis in the 2020s has also had a major effect. London is no longer dependent on external energy supplies, and those long-held objections to local energy stations have been tempered by the domestic energy production modules that most buildings now have. London has more parks, green space and wildlife than many rural areas, as streets and roads have been planted with trees and shrubs in the place of cars and traffic signs. The dramatic reduction in food consumption in the 2020s when high sugar, fatty and processed foods were banned has had the same effect on health as earlier bans on tobacco and alcohol. London may be a larger and denser city but is now more self-sustaining than it has ever been.

So, what next for London? Will 2062 herald a new swinging sixties era? Many of today's most active people were born in the 1960s and are the product of that generation. What has London learned in this last century? That change is inevitable, and London's energy to lead its own future is not diminished. London is now at the heart of the USE and a leading member of the federated state of Great Britain. Only the problem of England remains.

# A despatch from the future

## Matthew Gandy

Writing in the spring of 2062 the turmoil of the 2040s should have come as no surprise. The death of King William V in a skiing accident (only one resort remained in Europe because of climate change) led to the accession in 2037 of the now heavily bloated 'playboy prince' Harry to the UK throne (consisting of England and Wales after Scottish independence in 2024 and Irish re-unification in 2030). The steady stream of revelations, including extensive tax avoidance and money laundering, led to public demands for a referendum on the future of the monarchy which was narrowly passed in 2040 but then overruled by a obscure legal move instigated by the 'Bullingdon' faction of the re-elected New Conservatives. Tensions were already running high following the mysterious breakdown of the computerized voting system (outsourced to Belize) during the tightly fought general election of 2039. Following a much delayed and rain drenched Robbie Williams comeback concert in Hyde Park in 2041 (now aged 88) a vast crowd had attempted to storm the now permanently cordoned off 'district 3' created from an amalgamation of London's richest boroughs following the local government reorganization of 2028 that also saw the city's metropolitan boundary extended to the M25 orbital (now doubled in width in both directions). The overstretched and underpaid private security firm Peel, in charge of London's policing, had used live rounds on the irate crowd as they began to storm the high-end Hyde Park 2 residential towers. Following the disturbances in west London a series of security zones in other UK cities were also stormed, along with several police stations, now re-named 'security control points', in Leeds, Manchester and the vast 'Medway super city' (an elongated London overspill zone stretching from Gravesend to Whitstable). Particular fury was vented at these 'control points' because they managed the use of facial algorithms and DNA barcoding to restrict access to banks, hospitals, shopping centres and other buildings on a routine basis. The widespread use of 'electronic clamping', to render citizens 'inactive', and thereby excluded from society, had reached levels of 40 per cent by the early 2040s, with widening disparities in income and life expectancy across different parts of the city.

The eventual abandonment of the vast 'Thames Barrier 2' project in 2025, after the UK's third debt default, led to the eventual creation of a 'flood park' in 2035 stretching from the now derelict former Olympics site at Stratford to Rainham beyond the city's former metropolitan boundary.

**How to cite this book chapter:**
Gandy, M. 2013. A despatch from the future. In: Bell, S and Paskins, J. (eds.) *Imagining the Future City: London 2062*. Pp. 163-165. London: Ubiquity Press. DOI: http://dx.doi.org/10.5334/bag.y

The Thames Gateway development scheme initiated in the 1990s had been eventually dropped because of the withdrawal of insurance cover for new homes, leading to further pressure on the city's housing market, which contributed to the widespread 'shelter riots' of 2021, 2037 and especially 2042. The shortage of land for new housing had also been exacerbated by two further developments: the construction of a new international airport, with four runways, in Leytonstone, which had displaced over 200,000 people; and the growing trend for 'urban villas', constructed in a dizzying array of architectural styles on individual 1 hectare plots, sold principally to overseas buyers at 1 trillion dollars apiece.

The aerial view of London in 2062 is dominated by the green wedge-shaped expanse of its controlled flood zone beyond which we can observe a patchwork of giant fields in which all food production is managed by the food-healthcare conglomerate BupaFood Incorporated (the term 'food' itself had eventually been trademarked in 2031 following unexplained interruptions to the supply of basic commodities such as wheat, milk and soya). The now permanent presence of vast stretches of standing water in east London, following the disastrous flood of 2033, has had some unexpected consequences for the city's twelve million inhabitants: new strains of encephalitis, malaria and even dengue fever have become an everyday hazard for low-income populations, now almost exclusively concentrated in the heavily overcrowded and dilapidated 'non investment' zones. Although the word 'slum' is forbidden in all mass media outlets, now controlled by just one magnate, tattered copies of books by Mike Davis, David Harvey and other writers still circulate on the black market, in contravention of the 'digitization' edict of 2040, which sought to bring all forms of text-based communication under central control by Amazon IKL (Information, Knowledge and Leisure) based in New York and Shanghai.

London's manager, the term 'mayor' was dropped because of its democratic connotations, has been appointed on a quinquennial basis by the now enlarged Corporation of London since 2028, when they assumed full control of the metropolitan region for planning, security and 'policy delivery' (all services are now outsourced and the term 'public' is seen as highly anachronistic). Given the size and complexity of London the attempt by the Corporation and its sponsors to control all aspects of everyday life has its limits: a glance from the thickened glass windows of bullet trains, on the eventually completed Crossrail project, reveals the tell-tale signs of local food production and clandestine allotments nestling between abandoned buildings in the city's flood zone. Only the other day a neighbour gave me some ripe mangoes that had been secretly cultivated in our street.

# Reflections of a retiring bobby

## Aiden Sidebottom and Justin Kurland

Horizon-scanning has long had a chequered past, but nowhere more so than in policing. Many anticipated threats fail to materialise and many unforeseen crime opportunities are duly exploited. I, too, am no Nostradamus. On joining the Metropolitan Police Service in 2012, I had a vision of what the future of crime and policing in London would be like. Fifty years on, on the eve of my retirement – we retired younger back then! – it is amusing to reflect on how wrong I was.

I'll begin with cycle theft. Traditionally, cycle theft was considered a low police priority. Many stolen cycles were never reported to the police, few cycle thieves were caught and, frustratingly, many recovered bikes failed to reach their rightful owners, because most cyclists were unable to provide sufficient proof that they owned the bike in the first place. Then came Boris. In an attempt to reverse the spiralling obesity rates and pollution levels at the time, the then Mayor of London (and latterly England's First Minister) embarked on a sustained effort to promote cycling as a healthy alternative to motorised transport. The pro-bicycle trend continued and cycling flourished, so much so that the London of today reminds me of the Amsterdam of yesteryear. If only London had heeded the maxim 'think thief'. Even then there was strong evidence that bicycle ownership is positively correlated with levels of cycle theft, but commensurate security measures to guard against theft increases, such as the provision of secure parking facilities and robust locks, came much later. Cycle theft grew considerably and remains the scourge of present-day London. As a cyclist myself, theft is an unfortunate occupational hazard. Re-cycling acquired a new meaning when theft victims would quietly visit places like Brick Lane market to acquire a replacement of uncertain provenance, and even occasionally their own stolen model.

Cycle theft is an example of how changes in London generated opportunities for crime. Commodity theft is an example of how opportunities for crime generated changes in London. The term 'commodity theft' did not even exist when I joined the force in 2012. Metal theft was admittedly a problem, particularly the theft of copper cabling from the railway network. The recurrent explanation was that increases in the price of metals, in response to global demand exceeding supply, increased the profitability of stealing metals. That imbalance was never redressed. Thus came the now much-storied spates of petrol thefts (2020s), aluminium thefts (2030s) and so on. At the

**How to cite this book chapter:**
Sidebottom, A and Kurland, J. 2013. Reflections of a retiring bobby. In: Bell, S and Paskins, J. (eds.) *Imagining the Future City: London 2062*. Pp. 167-168. London: Ubiquity Press. DOI: http://dx.doi.org/10.5334/bag.z

time, identifying cost-efficient, functional alternatives to the widely-used (and widely stolen) metals and fuels was, excuse the pun, a pipe dream. But as human society repeatedly shows, science and engineering found a way. London is now awash with fibreglass manhole covers, fibre-optic cabling, etc, that stand in stark contrast to their metal-clad forebears. A large part of this change can be attributed to the debilitating effects of commodity theft on London's infrastructure. And most importantly to me, the change meant that opportunities for commodity theft were rendered (virtually) obsolete, with little displacement to the non-metallic substitutes.

We used to have a saying in the police, 'oh to be young, and to feel the *cops'* keen sting' (the author and correct wording escapes me). There was some truth to the statement in my formative years, with recorded crime statistics confirming that those aged 16 – 24 both perpetrated and experienced disproportionately higher rates of crime. But London then was a young city. Now it is old. I say that not as a long-in-the-tooth sourpuss—which I admittedly am—but on the basis of evidence. London, like many industrialised cities, witnessed huge shifts in the age structure of its population, facilitated largely by myriad improvements in medical science. Crime changed too. Most obviously, crime on the whole reduced, as the proportion of London's population became older (cycle theft excluded). Moreover, scanning through last year's recorded crime figures for London, I see that there are more over-65s in Brixton prison than there are 18 – 24 year olds. If I had quoted these figures in 2012, my colleagues surely would have laughed at my expense, believing instead that I had confused a prison roll call with that of a retirement home.

And the single greatest change? Not crime, but the way we respond to it. I joined the Met at a time when so-called *predictive policing* was gathering momentum. The principle was sound: research evidence consistently demonstrates (it still does!) that prior victimisation is the best predictor of future victimisation, both to the victim and to comparable targets nearby. Elevated risk gradually diminishes over time, however. Predictive policing harnessed these recurrent patterns, using sophisticated computer algorithms to identify the geographic areas where, probabilistically on the basis of previous evidence, crime is most likely to occur. Early demonstration projects in Manchester and Los Angeles were promising, and were later confirmed by similar efforts internationally. What was fledgling is now commonplace. What was predictive policing is now, simply, policing. And the London Met is arguably its most adept practitioner. For the last decade or so, the lion's share of decisions regarding the deployment of police resources and personnel in the capital has been informed by PRECRIM (PREdictive Crime Risk Models); the Met's supercomputer named to honour researchers at the University College London Jill Dando Institute, who were pioneers in the application of predictive policing. Implementation was slow and resistance was staunch, with many senior police officers viewing the innovation as an insult to their expertise and discretion. The resistance proved unfounded. The growth of predictive policing no more did away with police officers than the advent of laser-optic surgery did away with surgeons. The relationship is mutually beneficial. The tipping point was performance. The police, rightly or wrongly, have always been measured on their performance. We perform better with PRECRIM. The change? After decades of attempts and exhortation, science and the scientific method has finally replaced the tyranny of tradition, divination, hunches and folk wisdom. And London is better for it.

# London after London

## Matthew Beaumont

'I have no more faith than a grain of mustard seed in the future history of "civilization", which I *know* now is doomed to destruction, and probably before very long,' wrote the socialist William Morris in 1885. 'What a joy it is,' he added in a tone of vengeful satisfaction, 'to think of barbarism once more flooding the world, and real feelings and passions, however rudimentary, taking the place of our wretched hypocrisies...'

Morris's deliciously intemperate outburst was inspired by *After London* (1885), a curious novel by the naturalist Richard Jefferies, which depicts a society reshaped by some nameless environmental cataclysm that has almost completely destroyed the British capital. 'This marvellous city, of which such legends are related,' its narrator unsentimentally comments, 'was after all only of brick, and when the ivy grew over and trees and shrubs sprang up, and, lastly, the waters underneath burst in, this huge metropolis was soon overthrown.'

Morris was bewitched by this vision. 'Absurd hopes curled round my heart as I read it,' he confided; 'I wish I were thirty years younger. I want to see the game played out.' In a sense, Morris himself played the game out—in the form of *News from Nowhere* (1890-91), a utopian romance that he set in London approximately a century after a socialist revolution that, in 1952, transformed the nation's social relations. But, compared to *After London*, there is something rather tame about Morris's pastoral vision of twenty-first century England, which is centred on a series of picturesque descriptions of Bloomsbury. It domesticates Jefferies's fanaticism.

Like Morris, I too have always found the descriptions of the city's destruction that Jefferies included in *After London* deeply seductive. Indeed, when I see disaster movies in the cinema, I secretly dream about barbarism once more flooding the world. Absurd hopes curl round my heart as I watch entire cities being reduced to rubble by biblical floods and quakes. Because in these ends there is a beginning. The silence of almost empty streets, filmed at first light by the directors of contemporary disaster movies, before the roads and pavements are convulsed by the rhythms of commuters, trembles with utopian promise. 'The post-catastrophe situation,' in Fredric Jameson's formulation, 'constitutes the preparation for the emergence of Utopia itself.'

**How to cite this book chapter:**
Beaumont, M. 2013. London after London. In: Bell, S and Paskins, J. (eds.) *Imagining the Future City: London 2062*. Pp. 169-172. London: Ubiquity Press. DOI: http://dx.doi.org/10.5334/bag.aa

London in 2062 looks like the scene from the archetypal disaster movie in which, in that post-coital calm after the calamity itself, it suddenly becomes obvious that civilisation will have to be totally rebuilt if it is to be saved at all. It is a post-apocalyptic city, one that subsists in the aftermath of an environmental collapse that has rendered the fragile distinction between developed and developing nations quite meaningless. From a distance, it is impossible to decide whether, back in 2012, this city was London, Mumbai or Nairobi. Certainly, the survivors are people of all colours and creeds, and they don't seem to care about their differences.

The city has been scorched and scarified by some unfathomable human accident, possibly caused by a nuclear disaster—though already nobody can exactly remember the first cause. It has been set adrift by floods; honeycombed by bombs; and hurriedly tunnelled out by looters. London's financial district has collapsed into its foundations, and an immense population of computers, chittering like rats, softly decomposes amidst the dust of buildings. 'London, Paris, Tokyo, New York,' reads one graffito on the ruined façade of a department store: 'Fission Capitals of the World...'

But centuries of creeping sickness have at least come to an end in these unthinkable seizures, and the city finally seems to be convalescent. Far beneath its pavements lie the fields, forests and streams on which its health had once been founded. A process of forestation is in fact already silently taking place in the metropolis. Emerald-coloured vegetation protrudes with gentle insistence through the cracks that pave the streets. Lichen silently and relentlessly colonizes ancient road signs, like the *Xanthoria Parietina* that prompts the eponymous character of Patrick Keiller's film *Robinson in Ruins* to affirm his belief in 'a network of non-human intelligences' which are 'determined to preserve the possibility of life's survival on the planet.' There are even rumours that it is sometimes possible to glimpse zoo animals that, miraculously, have managed to survive in the city in spite of its devastation.

At mealtimes, people collect in the public parks, remnants of the past that suddenly resemble maps of a different future. The people picnic in craters on plants that have been carefully harvested from the cracks and crevices that vein the surrounding roads. Overlooking these scenes, cente-

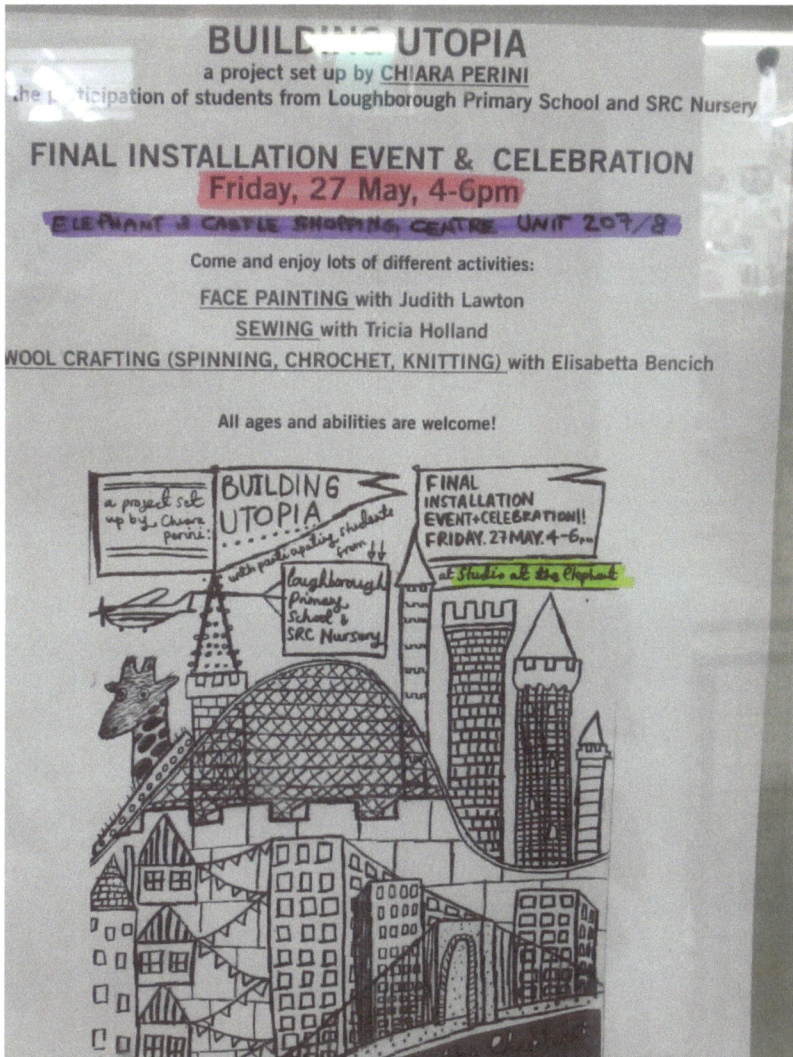

narians recall their long deceased parents' magical descriptions of playing in the ruins of London during the decade after the German air raids, when Rosebay Willowherb, which everybody called 'bombweed', thrived in the cavities and declivities opened up by the shelling. In the 'nonlinear dimensions of urban ecology', to paraphrase Mike Davis, a nonlinear model of history, radiating and proliferating like the ruderals beneath the roads, can faintly be discerned.

The city is excavating itself; it is reconstructing itself. 'The very bones of the city, all her history, from Roman times to the present', are in Michael Moorcock's words 'exposed and clearly visible'. And, among these bones, some of them ancient, people are reimagining the city's musculature. Citizens of all ages and backgrounds are collectively beginning to outline architectural blueprints for fully socialized housing projects, ones that might eventually fulfil the dreams of the twentieth-century avant-garde.

Human beings are starting to rediscover the forms of a communal life they had long forgotten. In spite of the devastation, the city's reclaimed spaces pulse with some dim hope. People collect in the dusty transepts of disused churches or the half-flooded atriums of deliquescent hotels. There they argue, passionately, patiently, about how best to rebuild the metropolis and transform this city of the dead into the city of the future. Real feelings and passions, in Morris's phrase, are finally replacing the hypocrisies we have lived in the name of civilization.

# It's Malaria Season

## Mosquitos breed in the tube tunnels.

## Take your medication before entering zone 1

For live outbreak information:
http://londinium.eurorail.com/malaria
@LdnMalariaNews

Working together:

**euroRail**
Londinium

**AFRIAID**
AFRICAN AID FOR
EUROPE AND
NORTH AMERICA

**Mayor of**
**LONDINIUM**
By appointment of
H.R.H King George VII

**NHC** Ltd.
Voted England-Wales
best health company
2062

# Conclusions

## James Paskins and Sarah Bell

Some questions are very easy to ask, but turn out to be very difficult to answer. Asking questions about the distant future of an entire city, home to over 8 million people, is a good example. It is relatively straightforward to ask how the city will change over the five decades between London 2012 and London 2062, but practically impossible to provide a definitive answer.

Some might even consider that it is foolhardy to try, and there is much fun to be had in holding up predictions from the past to scrutiny in the present. However, rather than offering up hostages to fortune, the aim of this book was to open up debate about the choices we can make now in support of the futures we wish to see. The preceding chapters are based on serious debate about the future of London, informed both by academic enquiry and professional experience. We feel that this book has taken an exciting approach by combining serious, structured, deliberation with more creative pieces. This combination reveals a range visions of the future that are, at turns, fun, fascinating and unsettling. Allowing clever, engaged and well-informed people the chance to follow their intuitions has proved to be a valuable and illuminating exercise.

One of the most challenging aspects of this exercise is the question of scope. London is a fantastically complicated entity. A city, especially a city as large and diverse as London, cannot be subdivided into neat, distinct categories. It certainly does not conform to the disciplinary structure of a university. So, what aspects of London are we interested in, and what level of detail should we cover? In large part this was answered by the contributors themselves, who have applied their knowledge to the question of the future of London, following their interests without overstepping the bounds of their expertise.

The choice of timeframe is another aspect of the scope. How long is fifty years? It is long enough to free people from thinking about the annual budgets and election cycles, long enough to expect to see a different world. However, much of the city, its buildings, its layout and its traditional institutions, may be left relatively untouched by the passage of five decades. This is one reason why retrofit might become such an important aspect in delivering a sustainable future for the city. Barring revolution, war and disaster, there is little reason to expect the tourist's view of London to change. The visitor's snapshots of St Paul's Cathedral, The Palace of Westminster and the Lord

**How to cite this book chapter:**
Paskins, J and Bell, S. 2013. Conclusions. In: Bell, S and Paskins, J. (eds.) *Imagining the Future City: London 2062*. Pp. 173-175. London: Ubiquity Press. DOI: http://dx.doi.org/10.5334/bag.ab

Mayor's Show, might be in glorious immersive 3D by 2062, but we might expect that the scenes they capture will be largely indistinguishable from today.

However, as we stare into the cloudy crystal ball of the future, we should not discount the possibility of radical change. Within the space of a week the inferno of 1666 reduced many of the capital's landmarks to ashes, it is not impossible that another 'Great' event will radically alter the city between now and 2062. While fire seems less likely, the resilience of the city may well be tested by the effects of climate change, and future history could record a 'Great Flood' or 'Great Swelter'. Nor should we ignore the spectre of terrorism. An increasing reliance on an expanding technology infrastructure exposes the city to another risk. If our future smart cities are underpinned by a tightly linked system, with inadequate protection or fail-safes, we could see critical systems drastically affected by a malicious act, or an accident of nature.

The development of smart cities infrastructure provides more than just one more vulnerability for the city, it also introduces the prospect that new technologies will change our relationships with, and understanding of, the city. It is likely that as the city gets smarter it will transform commerce, working practices and decision-making.

Cities exist for, and because of, people, and there is no doubt that fifty years is a significant amount of time for the people who live and work in the city. The years leading up to 2062 will see everyone involved in this book past retirement into old age, unless of course retirement becomes a relic of an earlier, more secure, time. In the opening chapter of this book we are told that London is 'defined by its population'. If we accept this, it follows that much of what makes London a global leader in areas as diverse as art, business and innovation, is its population, the flow of people, their energy and their ideas. London is a young, vibrant city, drawing on a talent from the rest of the UK, Europe and the world. It is clear that in some of our possible futures this population profile, and London's ability to act as a magnet for talent, is under threat. This is not to suggest that the capital should exclude its older members; like much of Western Europe, London needs to seriously consider how it can be more inclusive across the age spectrum. The size and structure of London's population in 2062 will be a fundamental driver of decision-making and a major influence on quality of life.

We should not take for granted the factors that attract the people who make the capital successful. There is no simple formula for what makes a city attractive, but the factors must surely include opportunities for employment, education and entertainment, and a city that accommodates and becomes enriched by different cultures and lifestyles. An intolerant city, or one that fails to provide affordable housing, rewarding jobs, or loses sight of the people at the bottom of society runs the risk of becoming sterile and moribund.

Another strength is the diversity of opportunity and endeavour within London. Finance is often held up as London's defining industry, and it is of vital importance to the prosperity of the city, as well as its place on the world stage. But London also hosts a remarkable breadth of other industries that add to the attractiveness of the city for doing business. While it is important to support the interests of business in London, the interests of one sector should not be allowed to overwhelm other needs. Climate change, poverty reduction and wellbeing are all considerations that should be balanced against maintaining London's status as a global city.

It is clear that London's leaders must face up to some bold decisions in the coming decades. It seems clear that the next fifty years will bring different priorities into conflict. We face decisions on long-term infrastructure, including rail links, airports and tunnels under the Thames that will have direct impacts on our ability to meet targets on air quality, water pollution and carbon emissions. In many cases, such as transport, planning and energy policy, the need to maintain business as usual is in direct opposition to measures required to address human induced climate change. If we are to deliver a sustainable future, there is a pressing need for well-targeted investment, strong legislation and good ideas.

We have not, nor could we, cover every facet of London, neither were we able consider every possible future. However, the areas that receive coverage are not only important in themselves, but also for the questions they raise. And it is pressing questions, rather than answers, that we are left with. How will London provide for its citizens in the future? How many of them should we expect to see in 2062, and who will they be? It is clear that these people will still require shelter, water and energy. It also seems clear that London will rely on its connections to provide not only these staples, but also the flows of energy and resources required to meet the aspirations of its citizens. There are questions of governance and representation, finance and fairness, business and society. All of these questions, and many more will benefit from cross-disciplinary and cross-sector approaches, bringing together diverse expertise to generate new insights and future directions.

So, what will the future hold for London 2062? We may be the citizens of football clubs, perhaps we will spend our leisure time singing instead of shopping, and maybe a smart city running on big data will have changed our lifestyles beyond recognition. It is likely that London 2062 will be a different, and perhaps more dangerous, place. In *Dreams* we are warned to take sensible precautions before we consider travelling on the tube of 2062. We should be just as careful as we begin our journey towards the future; the decisions we make, the areas we research and the causes we support over the next fifty years will shape the London we see in 2062.

# Index

www.ingramcontent.com/pod-product-compliance
Lightning Source LLC
Chambersburg PA
CBHW041547260326
41914CB00016B/1580

* 9 7 8 1 9 0 9 1 8 8 1 8 1 *